ENERGY SCIENCE, ENGINEERING

CLEAN AND RENEWABLE ENERGY STANDARDS

OPTIONS AND OBJECTIVES

ENERGY SCIENCE, ENGINEERING AND TECHNOLOGY

Additional books in this series can be found on Nova's website under the Series tab.

Additional E-books in this series can be found on Nova's website under the E-books tab.

ENERGY SCIENCE, ENGINEERING AND TECHNOLOGY

CLEAN AND RENEWABLE ENERGY STANDARDS

OPTIONS AND OBJECTIVES

BRIAN J. RUTHER

AND

JACLYN R. MORAN

EDITORS

Nova Science Publishers, Inc.

New York

For permission to use material from this book please contact us:
Telephone 631-231-7269; Fax 631-231-8175
Web Site: http://www.novapublishers.com

Library of Congress Cataloging-in-Publication Data

Clean and renewable energy standards : options and objectives / editors, Brian J. Ruther and Jaclyn R. Moran.
    p. cm.
Includes bibliographical references and index.
ISBN 978-1-61324-932-1 (softcover : alk. paper) 1. Renewable energy sources--Standards--United States. I. Ruther, Brian J. II. Moran, Jaclyn R.
TJ807.9.U6C438 2011
333.79'4--dc23
                              2011018550

Published by Nova Science Publishers, Inc. † New York

CONTENTS

PREFACE

During his January 25, 2011 State of the Union speech, President Obama proposed a Clean Energy Standard (CES) policy framework that would result in 80% of U.S. electricity generation coming from "clean energy" sources by 2035. "Clean energy," as described by President Obama, would include renewables, nuclear power and partial credits for clean coal and efficient natural gas. This book examines the design elements of previous CES proposals, summarizes the Administration's CES policy framework, provides state-level baseline CES compliance analysis, and presents several policy options that Congress might consider as part of a CES debate.

Chapter 1- During his State of the Union speech on January 25, 2011, President Obama announced an energy goal for the country: "By 2035, 80% of America's electricity will come from clean energy sources." The White House, on February 3, 2011, released a Clean Energy Standard (CES) framework focused on U.S. electricity generation. The framework describes the fundamental goals and objectives of such a policy to include doubling clean electricity, sustaining and creating jobs, and driving clean energy innovation.

Chapter 2- The choice of power generation technology in the United States is heavily influenced by the cost of fuel. Historically, the use of fossil fuels has provided some of the lowest prices for generating electricity. But growing concerns over greenhouse gas emissions and other environmental costs associated with burning fossil fuels are leading some utilities and energy providers to deploy more renewable energy technologies to meet power demands.

Chapter 3- A global race is underway to develop and manufacture clean energy technologies, and we are competing with other countries that are playing to win. America has the most dynamic economy in the world, but we can't expect to win the future by standing still. That's why, in his State of the Union address, President Obama proposed an ambitious but achievable goal of generating 80 percent of the Nation's electricity from clean energy sources by 2035. Meeting that target will position the United States as a global leader in developing and manufacturing cutting-edge clean energy technologies. It will ensure continued growth in the renewable energy sector, building on the progress made in recent years. And it will spur innovation and investment in our nation's energy infrastructure, catalyzing economic growth and creating American jobs.

Chapter 4- In his recent State of the Union address, President Obama proposed a Clean Energy Standard (CES) to require that 80 percent of the nation's electricity come from clean energy technologies by 2035. The Senate Energy and Natural Resources (ENR) Committee now faces a threshold question of what the general policy goals for the electric sector are and whether a CES would most effectively achieve them. Is the goal to reduce greenhouse gas emissions, lower electricity costs, spur utilization of particular assets, diversify supply, or some combination thereof? Depending on the goals, is a CES the right policy for the nation at this time? If so, is 80 percent by 2035 the right target? If not, should alternatives to reach similar goals be considered?

Chapter 5- A Renewable Portfolio Standard (RPS) or Energy Standard (RES) requires a percent of energy sales (MWh) or installed capacity (MW) to come from renewable resources. Percents usually increase incrementally from a base year to a later target. The map on the front shows ultimate targets.

Chapter 6- 16 states and D.C. have created solar or distributed generation (DG) set-asides in their Renewable Portfolio Standards (RPS), to encourage development of higher-cost technologies so they can move closer to cost parity with other renewable resources.

In: Clean and Renewable Energy Standards   ISBN: 978-1-61324-932-1
Editors: B. J. Ruther, J. R. Moran   ©2012 Nova Science Publishers, Inc.

Chapter 1

CLEAN ENERGY STANDARD: DESIGN ELEMENTS, STATE BASELINE COMPLIANCE AND POLICY CONSIDERATIONS[*]

SUMMARY

During his State of the Union speech on January 25, 2011, President Obama announced an energy goal for the country: "By 2035, 80% of America's electricity will come from clean energy sources." The White House, on February 3, 2011, released a Clean Energy Standard (CES) framework focused on U.S. electricity generation. The framework describes the fundamental goals and objectives of such a policy to include doubling clean electricity, sustaining and creating jobs, and driving clean energy innovation.

Congress, if it chooses to take up CES legislation, will likely sort through and evaluate a number of policy options that might be considered during the formulation of a federal Clean Energy Standard policy. Understanding previous CES proposals, the Administration's CES policy framework, state-level baseline CES compliance, and policy considerations might assist a CES debate during the 112th Congress. These areas are the focus of this report.

---

[*] This is an edited, reformatted and augmented version of a Congressional Research Service publication, CRS Report for Congress R41720, from www.crs.gov, dated March 25, 2011.

CES and related concepts have been debated for more than a decade and several Clean/Renewable Energy Standard proposals were offered during the 111th Congress, although none became law. The scope of this report includes a comparative analysis of four proposals of the 111th Congress: S. 20, Clean Energy Standard Act of 2010; S. 3464, Practical Energy and Climate Plan Act of 2010; S. 3813, Renewable Electricity Promotion Act of 2010; and a substitute amendment offered for H.R. 2454, American Clean Energy and Security Act of 2009. This analysis, which illustrates commonality and key differences among the legislative proposals, includes an assessment of each bill based on a uniform set of design elements. While the proposals considered generally agree on the definition of "renewable energy" (wind, solar, geothermal, etc.), they differ on certain policy aspects including (1) base quantities of electricity, (2) target/goal for the standard, and (3) alternative compliance payments, among others.

The Administration's proposal states that 40% of delivered electricity is generated from "clean energy" sources today and 80% should be generated from clean energy sources by 2035. Clean energy sources are defined to include (1) renewable energy, (2) nuclear power, and (3) partial credits for clean coal and efficient natural gas. However, the amount of partial credits received by clean coal and efficient natural gas generation is not explicitly defined.

CRS analysis of 2009 electricity generation data from the Energy Information Administration (EIA) also suggested that 40% of electricity generated could considered clean energy if renewable energy, nuclear power, and 50% of electricity generated from natural gas combined cycle (NGCC) power plants are classified as clean energy. Further analysis of EIA data assessed the amount of clean energy generation in each state. This work revealed differences among the states regarding existing clean energy generation, with some states currently generating more than 80% of electricity from such clean energy sources and other states generating less than 5%.

Finally, the Clean Energy Standard debate involves several policy design options that Congress might consider, including (1) Should the policy credit existing and/or incremental clean energy generation? (2) What should be the value of alternative compliance payments? (3) Should utility companies of a certain size be exempt? (4) Should preference be given to renewable energy generation? and (5) Which generation sources would qualify as clean energy? These, and other, policy options are presented and discussed in this report.

BACKGROUND AND INTRODUCTION

In 2009, approximately 4 trillion kilowatt hours of electricity were generated by the U.S. power sector. By 2035 electricity generation is expected to rise to more than 5 trillion kilowatt hours, a roughly 25% increase from 2009 levels. The fuel mix for U.S. electricity generation includes four primary categories: (1) coal, (2) natural gas, (3) renewables, and (4) nuclear. As illustrated in Figure 1, coal is the largest electricity generation fuel source for both actual (2009) and projected (2035) generation. However, EIA projects that natural gas and renewables are the only fuel sources that would experience growth, in terms of percentage of the electricity generation mix, over the projection period.

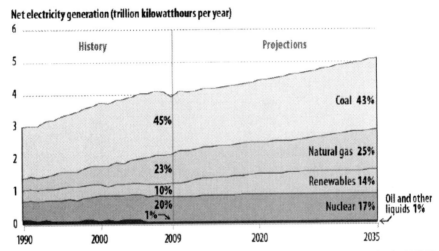

Source: Energy Information Administration (EIA), Annual Energy Outlook (AEO) 2011 Early Release Overview, available at http://www.eia.gov/forecasts/aeo/pdf/0383er(2011).pdf. Adapted by CRS.

Notes: EIA analysis assumes that current laws and regulations remain unchanged throughout the projection period. "Oil and other liquids" is a fifth fuel source category, however electricity generated from these fuel sources is marginal.

Figure 1. Projected Fuel Mix for Electricity Generation.

During his January 25, 2011 State of the Union speech, President Obama proposed a Clean Energy Standard (CES) policy framework that would result in 80% of U.S. electricity generation coming from "clean energy" sources by 2035. "Clean energy," as described by President Obama, would include renewables, nuclear power, and partial credits for clean coal and efficient natural gas.

While there is no official definition, a federal Clean Energy Standard might be defined as a requirement to generate a percentage of electricity from certain energy sources. It is a policy designed to encourage U.S. electricity generation from "clean" or "cleaner" energy sources within a certain time period. Many CES proposals require individual utility companies to comply with a federal CES, although some utilities may be exempt from CES requirements based on their total amount of annual electricity sales. Generally, utilities can comply with CES requirements through a combination of (1) electricity generation from qualified clean energy sources, (2) purchasing clean energy credits, and (3) making alternative compliance payments (ACPs). Each of these will be discussed below.

Previous CES proposals have addressed multiple policy design parameters, including (1) technologies that qualify, (2) base quantities of electricity, (3) goals and requirements, (4) alternative compliance payments. Understanding the implications and inter-relationships of these parameters is an important element of CES policy design and will assist Congress with considering if overall objectives such as increasing clean energy generation, minimizing rate payer impacts, and job creation are likely to be achieved.

This report evaluates design elements of previous CES proposals, summarizes the Administration's CES policy framework, provides state-level baseline CES compliance analysis, and presents several policy options that Congress might consider as part of a CES debate.

SUMMARY AND DESIGN ELEMENTS OF PREVIOUS CLEAN ENERGY STANDARD PROPOSALS

During the 111th Congress several Clean/Renewable[1] Energy Standard policies were proposed, although none became law. In order to provide background on previously proposed CES legislation, four proposals were analyzed and compared against multiple design parameters (See Table 1).[2]

Table 1. Clean/Renewable Energy
Standard Proposals—111th Congress

Bill #	Title	Proposed Standard
S. 20	Clean Energy Standard Act of 2010	50% of base quantity by 2050.
S. 3464	Practical Energy and Climate Plan Act of 2010	50% of base quantity by 2050.
S. 3813	Renewable Electricity Promotion Act of 2010	15% of base quantity by 2021.
Substitute amendment to H.R. 2454	Energy Production, Innovation, and Conservation Act	15% of base quantity by 2020.

Source: Legislative Information System and Congressional Quarterly.

Notes: S. 3813 might be considered a purely "Renewable" Energy Standard as it does not include nuclear or other fossil energy generation as qualified sources. The substitute amendment to H.R. 2454 includes a number of energy policies and the only portion analyzed was Title 1—Clean Energy Standard. The substitute amendment is dated May 19, 2009. There may have been a subsequent version with some changes/modifications submitted to a committee. S. 3464 also includes several other energy policies and this analysis focused on Title III—Diverse Domestic Power. Each proposed standard analyzed has a different definition for "base quantity" of electricity.

A more detailed analysis of these proposals is presented in a side-by-side comparison matrix that can be found in Appendix A. All proposals were compared in order to assess areas of commonality and divergence. While not an all-inclusive or exhaustive list, following is a brief overview and discussion of the design elements considered for this analysis.

Base Quantity of Electricity

The base quantity of electricity is a critical Clean Energy Standard design element as it establishes the amount of electricity, typically measured in kilowatt-hours, that applies to CES goals and requirements. Proposals analyzed have base quantity definitions that range from 100% of utility power sales to sales less the amount of power generated by hydro-electricity and municipal solid waste (MSW) incineration. A hypothetical example of how different utilities might derive their respective CES base quantities in the latter case is provided in Table 2.

Table 2. Hypothetical Base Quantity Calculations

Pre-Adjustment		Factors for Adjustment		Pre-Adjustment Less Factors	15% Standard
Total Sales (kWhrs)		Hydroelectricity (kWhrs) MSW (kWhrs)		Base Quantity (kWhrs)	CES Target (kWhrs)
Utility 1	100 billion	0	0	100 billion	15 billion
Utility 2	100 billion	50 billion	50 billion	0	0
Utility 3	200 billion	100 billion	50 billion	50 billion	7.5 billion

Source: CRS

Notes: This analysis assumes that the base quantity of electricity is calculated by subtracting electricity generated from hydroelectricity and Municipal Solid Waste (MSW) incineration from the total amount of electricity sales.

Target/Goal

Clean Energy Standard targets and goals set the percentage of electricity that must be generated from clean energy sources by a certain date. The percentage articulated in a CES proposal is applied to the base quantity of electricity to calculate the number of kilowatt-hours that must be generated from clean energy sources in order to achieve compliance by a certain date. Examples of CES targets/goals include (1) 50% of base quantity by 2050, (2) 15% of base quantity by 2039, and (3) 15% of base quantity by 2020.

Qualifying Energy Sources

Defining and determining which energy sources will qualify under a CES proposal could be a design element worthy of consideration. A clear definition of qualifying sources is important as it allows a utility company to determine which electricity generation options are available for compliance. Each of the four proposals analyzed in this paper include typical renewable energy[3] as qualified sources. All four proposals also include coal-mine methane and landfill gas as qualifying sources. Differences among the proposals, generally, are associated with the inclusion and definition of qualified hydropower and incremental geothermal as well as the inclusion/definition of waste-to-energy, qualified nuclear, advanced coal/fossil with carbon capture and storage, and re-powering/co-firing biomass at existing coal generation facilities.

Energy Efficiency/Savings Credits

Energy efficiency/savings typically refers to reductions in electricity consumption at end-use consumer facilities that are served by an electric utility company as well as reductions in distribution system losses. Some proposals also include output from combined heat and power systems as energy efficiency/savings. In order to qualify for energy efficiency/savings credits, utility companies may have to institute programs that result in consumer demand reductions. One example of such a program might be subsidies for high efficiency air conditioning systems. All four proposals analyzed allow for energy efficiency/savings credits, although some proposals place limits on how much energy efficiency/savings credits can be used to comply with a broader Clean Energy Standard. For example, one proposal allows utility companies to use energy efficiency credits to satisfy up to 25% of the CES target. Therefore, 75% of a utility company's CES target must be met by generating electricity from qualified sources, purchasing CES credits, or making alternative compliance payments. One challenge associated with energy efficiency/savings credits might be determining a baseline for calculating energy efficiency and therefore the number of credits that result from various energy efficiency programs.

Alternative Compliance Payments

Alternative compliance payments (ACPs) can be paid by utility companies in lieu of generating qualified clean energy or purchasing clean energy credits. Typically expressed in cents per kilowatt-hour, ACPs provide utility companies with some degree of flexibility associated with meeting the targets/goals of a Clean Energy Standard. From a policy perspective, determining the value of ACPs can be somewhat complicated. Setting the ACP too low could potentially result in minimal development of "clean" electricity generation because some companies might choose to pay the ACP instead of generating or purchasing qualified clean energy. At the same time rate payer costs may increase as utility companies seek to recover their compliance costs. However, setting the ACP value too high might result in relatively large electricity rate increases in areas/regions that lack clean energy resources. Nevertheless, ACPs are basically a cost containment mechanism that effectively place a cap on the value of clean energy credits. Determining the value of ACPs will likely involve comparing the cost of generation from all qualified sources to the lowest generation cost from any fuel (e.g., coal, natural gas, nuclear, renewable) source. Some proposals suggest that funds generated through receipt of ACPs will be used to provide grants in support of new "clean energy" electricity generation projects.

Credit Trading

Under most of the four Clean Energy Standard proposals, utilities would be awarded credits for each kilowatt-hour of electricity generated from qualified clean energy sources. Utility companies can submit clean energy credits as a means of compliance with annual CES requirements. If a utility has more CES credits than are required for a given year, the utility may either "bank" the excess clean energy credits for a certain period of time or the utility can trade the excess credits, in exchange for cash, to other utilities. For those proposals that allow energy efficiency to count towards CES compliance, energy efficiency credits are typically handled in a similar manner. Details regarding the mechanics of how CES credit trading may work are not clearly defined in the four proposals analyzed and responsibility for establishing trading programs is delegated to the Secretary of Energy.

Multiple Credits

Three of the four CES proposals analyzed include provisions for double and triple credits. Multiple credits could be an approach that further incentivizes certain types of clean energy projects or the development of projects in certain locations. Some examples of projects that might receive multiple credits include (1) projects on Indian lands, (2) on-site electricity generation, (3) first five advanced coal facilities that sequester 1 million tons per year of carbon dioxide (CO_2), among others.

Credits for Demonstration Projects

Some CES proposals include provisions that allow demonstration projects to receive clean energy credits. For example, S. 20 would provide clean energy credits for advanced coal demonstration projects, based on the amount of CO_2 that is captured and sequestered. Providing credits for demonstration projects might be viewed as an incentive to develop, deploy, and commercialize emerging clean technologies.

Civil Penalties

Most CES proposals include a civil penalty for utilities that fail to comply with CES requirements. Civil penalties are typically computed by multiplying the annual kilowatt-hour target shortfall times a multiple of the alternative compliance payment (e.g., 200% of the ACP— inflation adjusted).

Exemptions

In some cases, CES proposals may exempt certain utility companies from compliance. Two of the four proposals analyzed exempt utilities that sell less than four million megawatt[4]-hours of electricity in the preceding year. All utility companies in Hawaii are also exempt.

Loans to Support Compliance

Some CES proposals empower the Secretary of Energy to make loans to support the development of qualified clean energy projects. The purpose of the loans is to assist with CES compliance and reduce cost impacts to utilities and retail consumers.

PRESIDENT OBAMA'S CLEAN
ENERGY STANDARD PROPOSAL

President Obama's Proposal for a Clean Energy Standard

Doubling the share of clean electricity over the next 25 years. To mobilize capital and provide a strong signal for innovation in the energy sector, a CES should be established that steadily increases the share of delivered electricity generated from clean energy sources, rising from 40% today to 80% by 2035.

Credit a broad range of clean energy sources. To ensure broad deployment and provide maximum flexibility in meeting the target, clean energy credits should be issued for electricity generated from renewable and nuclear power; with partial credits given for clean coal and efficient natural gas.

Protecting consumers against rising energy bills. The CES should be tailored to protect consumers, and coupled with smart policies that will help American families and businesses save money by saving energy.

- The CES should be paired with energy efficiency programs that will lower consumers' energy bills, such as stronger appliance efficiency standards, tax credits for energy efficiency upgrades, and the proposed Home Star program.
- The CES should also include provisions to help manufacturers invest in technologies to improve efficiency and reduce energy costs.

Ensuring fairness among regions. Different regions of the country rely on diverse energy sources today, and have varying clean energy resources for the future. The CES must ensure that these differences are taken into account, both among regions and between rural and urban areas.

Promoting new technologies such as clean coal. The CES should include provisions to encourage deployment of new and emerging clean energy technologies, such as coal with carbon capture and sequestration.

- Source: White House Office of Media Affairs, "President Obama's Plan to Win the Future by Producing More Electricity Through Clean Energy," February 3, 2011, available at http://www.whitehouse.gov/the-press-office/2011/02/03/president-obama-s-plan-win-future-making-american-businesses-more-energy.

On January 25, 2011, during the State of the Union address, President Obama announced a clean energy goal for the country: "By 2035, 80% of America's electricity will come from clean energy sources."[5] On February 3, 2011, the White House released a document titled "President Obama's Plan to Win the Future by Producing More Electricity Through Clean Energy," which summarizes the goals of the President's plan.[6] Primary objectives of the Administration's plan include:

- Double the share of clean electricity in 25 years
- Draw on a wide range of clean energy sources
- Deploy capital investment to sustain and create jobs
- Drive innovation in clean energy technologies
- Complement the clean energy research and development agenda

Furthermore, President Obama's plan described five core principles for the Clean Energy Standard proposal. These principles are summarized in the following text box above.

As discussed and presented in the following sections, baseline compliance with President Obama's CES proposal differs among the states and several policy considerations may warrant further evaluation as the CES policy debate evolves.

BASELINE STATE COMPLIANCE ASSESSMENT (PRESIDENT OBAMA'S CES PROPOSAL)

President Obama's Clean Energy Standard proposal states that 40% of electricity currently generated nationwide comes from "clean energy" sources.[7] However, each state and each utility required to comply with a federal Clean Energy Standard has a unique electricity generation mix. The following figure shows how each state would currently comply with the CES proposal based on existing electricity generation from qualified "clean energy" sources. Data sources and the calculation methodology used to generate Figure 2 and Figure 3 are described in Appendix B.

As indicated in Figure 2, some states may be better positioned than others to comply with a Clean Energy Standard, with some states already exceeding the 80% goal for 2035 and other states generating a relatively small percentage of electricity from qualified clean energy sources. Previously proposed CES legislation has typically applied to electric utilities and has been based on the amount of electricity sold to consumers.

The same data used to create Figure 2 were used to create the map shown in Figure 3. This map illustrates potential regional differences associated with CES compliance, based on existing electricity generation sources. This map does not provide utility-level percentages, which could be the basis for CES implementation.

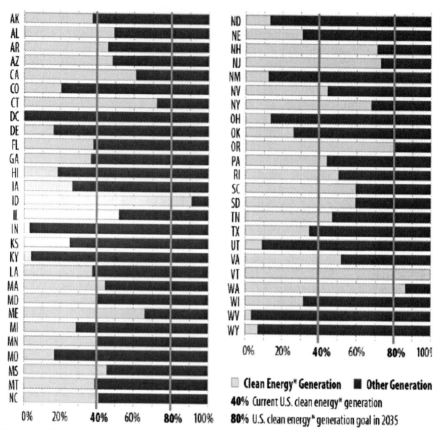

Source: CRS analysis of Energy Information Administration (EIA) Electric Power
    Annual 2009—Data Tables and Form 923 electricity generation survey data for
    2009.

Notes: * Sources classified as Clean Energy, based on the White House Clean Energy
    Standard framework, include geothermal, hydroelectric, nuclear, biomass, pumped
    storage, solar thermal and photovoltaic, wind, wood/wood derived fuels. Also, a
    50% CES credit per kilowatt hour was provided for natural gas combined cycle
    (NGCC) generation.

Figure 2. State-Level Clean Energy Compliance (2009 electricity generation).

Limits of Analysis

Information provided in this report does not provide specifics at the utility
level and does not represent the total amount of electricity sales to consumers.

Such level of analysis is beyond the scope of this report. Nevertheless, the information presented here does illustrate generation profile differences among the U.S. states and may be useful as a baseline assessment of state and regional differences associated with CES legislation.

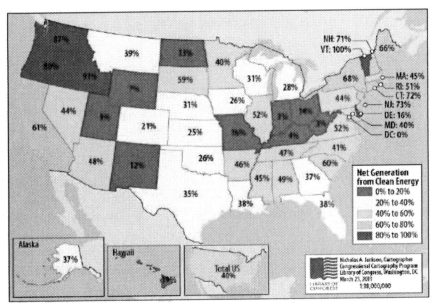

Source: CRS analysis of Energy Information Administration (EIA) Electric Power Annual 2009—Data Tables and Form 923 electricity generation survey data for 2009.

Figure 3. Regional Clean Energy Standard Compliance Assessment (2009).

Should the Policy Provide Credit to Existing and/or Incremental "Clean Energy" Generation?

President Obama's CES proposal states that 40% of U.S. electricity is generated from "clean energy" sources. However, as illustrated in Figure 2 and Figure 3, each state has a different generation mix that results in a wide range of initial baseline compliance levels. Allowing existing "clean energy" generation to count toward the standard would enable each state to receive credit for its respective "clean energy" capacity. However, allowing existing generation to count toward a CES puts some states in a better position when compared to other states, as indicated in Figure 3. Under this scenario, and depending on specifics of the proposed legislation, some states may experience some degree of wealth transfer as a result of purchasing CES credits from states with an excess of qualified "clean energy" electricity generation. Alternatively, Congress might decide to only allow incremental generation capacity added after the policy is enacted to count towards CES compliance. If such a policy were adopted, Congress may choose to structure the CES in a different manner than that proposed by President Obama. For instance, President Obama's 80% of total electricity generation by 2035 would be much more difficult to achieve if existing qualified generation sources are not eligible.

What Should Be the Value of Alternative Compliance Payments?

As discussed earlier, alternative compliance payments (ACPs) provide utility companies with some degree of flexibility associated with CES compliance. In essence, ACPs act as a cost ceiling for complying with a Clean Energy Standard. Setting the value of ACPs can be complicated by factors such as the cost of electricity generation, transmission availability, regional "clean energy" resources, and finance costs for advanced technology. As a result, setting a single ACP that encourages "clean energy" electricity generation for the entire country can be difficult and challenging. An ACP set too low may simply raise rate payer electricity costs and encourage minimum amount of "clean energy" generation. In contrast, an ACP set too high may not be acceptable for states that are not endowed with "clean energy" resources. Evaluating the levelized cost of electricity (LCOE) of qualified "clean energy" generation options may be a way to begin estimating an ACP.[8] However, since each region's "clean energy" resource base varies (solar in the southwest versus the northeast) and each technology may have different financial requirements due to real or perceived levels of technology risk, an LCOE-based analysis of ACP levels may, at best, only produce a reasonable range for the ACP. Setting a single, absolute ACP value that will be perceived as fair and equitable for all regions, and for all technologies, may be a challenging endeavor.

Should Utility Companies of a Certain Size Be Exempt?

Three of the four legislative proposals analyzed for this paper exempt certain utilities from complying with the respectively proposed Clean/

Renewable Energy Standard. Two proposals exempt utilities that sell less than four million megawatt hours and one proposal exempts utilities that sell less than one million megawatt hours. If Congress were to choose to exempt certain utilities from compliance with the proposed standard, an analysis of how much electricity generation is represented by exempt utilities as a percentage of total U.S. electricity generation may be useful. CRS analyzed EIA data to estimate two items: (1) the number of utility companies that would be required to comply with a CES, and (2) the amount of electricity sales represented by non-exempt utility companies. The analysis assumed that utility companies selling less than four million megawatt-hours per year are exempt. Results from this analysis are provided in Figure 4.

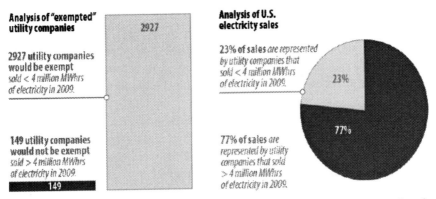

Source: CRS analysis of Energy Information Administration Form 861 survey data for 2009.

Notes: EIA Form 861 includes annual retail sales (in megawatt hours) information for more than 3,000 utility companies in the U.S. CRS categorized each utility based on annual retail sales being either more than or less than 4 million megawatt hours per year. Many of the 2,927 exempt utility companies include electric cooperatives and municipal utilities. Utility companies included in EIA's Form 861 survey data may include wholly owned subsidiaries of a parent company that may reflect retail sales less than 4 million megawatt hours. The parent company of these subsidiaries may have aggregate retail sales of more than 4 million megawatt hours. However, an assessment of parent and subsidiary companies is beyond the scope of this analysis. Please see the preceding paragraph for more detail on possible qualified exemptions. EIA's Form 861 data can be found at http://www.eia.doe.gov/cneaf/electricity

Figure 4. Analysis of Potential Utility Company Exemptions.

According to the analysis summary in Figure 4, 149 of more than 3,000 utility companies would have to comply with the CES based on the assumed exemption criteria. These 149 utility companies represent 77% of annual U.S. electricity sales. Including an exemption as part of a CES policy may prompt consideration how to effectively achieve a CES target given that a portion of U.S. electricity might be exempted from compliance. Based on the above analysis, non-exempt utilities could be required to generate more (greater than 80% by 2035 for example) electricity from "clean energy" sources in order to meet an 80%-by-2035 goal, assuming that was a goal established through legislation.

How Should Interim CES Targets/Goals Be Structured?

All four proposals from the 111th Congress include interim targets for CES implementation. These interim milestones serve as a means to phase in "clean energy" over a period of time. Figure 5 illustrates three possible phase-in approaches.

First, the linear approach, which might consist of annual increases, may be advantageous to renewable energy and natural gas generation since development timelines for these sources are relatively short. However, nuclear and "clean coal" may be at a disadvantage under this scenario due to long development timelines (nuclear) and technology maturity/commercialization ("clean coal").[9] Second, the back-end loaded approach, where targets are low in the beginning years of a policy and then increase steeply in later years, may be beneficial for nuclear and "clean-coal" generation as it allows more time for development and commercialization. However, under this scenario if beginning year targets are too low some may argue that this approach does not result in demand large enough to incentivize investment in new renewable and natural gas projects. Finally, the stepped approach might include targets and goals that increase every three to five years (example: 45% by 2015, 50% by 2020, etc.). This approach offers an alternative phase-in option but may result in flurried periods of project development followed by periods of stagnant, or non-existent, market growth. Manufacturing and job sustainability may be challenged under a stepped scenario.

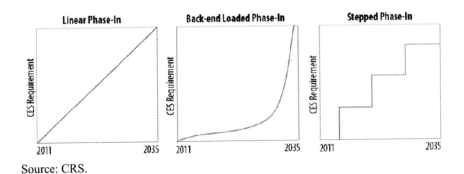

Source: CRS.

Figure 5. CES Phase-In Conceptual Approaches.

Should Preference Be Given to Renewable Energy Generation?

President Obama's proposal allows a number of "clean energy" generation sources to qualify for CES compliance. This approach provides utility companies with some degree of flexibility when choosing among "clean energy" generation alternatives and it allows nuclear, coal generation with carbon capture and storage, and natural gas to compete directly with renewable (wind, solar, geothermal, etc.) generation.

Some may advocate a preference for renewable energy in the form of a specific percentage of generation dedicated to renewable energy or through a multi-tiered CES credit approach that provides more credit value for electricity generated from renewable sources.[10] Others may argue that a CES should include a broad array of qualified electricity generation sources and state/regional markets should determine the generation mix selected for CES compliance.

Which Generation Sources Qualify as "Clean Energy"?

Qualified "clean energy" sources described in President Obama's CES proposal include (1) renewable electricity, (2) nuclear power, and (3) partial credits for clean coal and efficient natural gas. The proposal indicates that "clean coal" refers to coal-based electricity generation that includes carbon capture and sequestration and "efficient natural gas" refers to natural gas combined cycle (NGCC) electricity generation. Based on the choices of qualified sources, it appears that a "clean energy" objective is to encourage the development of low-carbon power sources. If this is the case, some may argue that supercritical and ultra-supercritical pulverized coal generation should qualify for partial credits since the carbon dioxide emissions profile is less than conventional subcritical pulverized coal generation. Sorting out qualified "clean energy" sources and determining the amount of whole and partial credits awarded for various electricity generation types could warrant further analysis in consideration of a federal Clean Energy Standard.

Other Policy Considerations

Other policy considerations may acquire increased levels of importance warranting further analysis and evaluation. Such issues include:

1) Transmission requirements and how to allocate associated costs?
2) Which federal agency should have responsibility for implementing and managing a federal CES?
3) What should be the guidelines for credit trading under a CES policy?
4) How might a federal CES affect other economic sectors, such as coal and coal electricity generation?
5) How should energy efficiency be treated under a federal CES?
6) How might a CES align and interact with renewable portfolio standards currently established in 29 states, DC, and Puerto Rico?

Senate Energy and Natural Resources: Clean Energy Standard White Paper

On March 21, 2011, the Senate Energy and Natural Resources committee released a Clean Energy Standard white paper. This white paper solicits feedback on 6 broad policy design questions along with 36 clarifying questions.[11] The six broad design questions listed in the white paper are (1) What should be the threshold for inclusion in the new program? (2) What resources should qualify as "clean energy"? (3) How should the crediting system and timetables be designed? (4) How will a CES affect deployment of specific technologies? (5) How should Alternative Compliance Payments, regional costs, and consumer protections be addressed? (6) How would a CES interact with other policies?

APPENDIX A. COMPARATIVE ANALYSIS OF SELECTED CLEAN ENERGY STANDARDS PROPOSED DURING THE 111ᵀᴴ CONGRESS

Table A-1. Clean Energy Standard Legislative Proposal Analysis (111th Congress)

	Clean Energy Standard Act of 2010 S. 20	Practical Energy and Climate Plan Act of 2010: Title III -Diverse Domestic Power S. 3464	Renewable Electricity Promotion Act of 2010 S. 3813	Substitute Amendment for H.R. 2454 (dated May 19, 2009)
Base quantity	Total electricity sold, less generation from hydroelectricity and MSW incineration.	Total electricity sold, less hydroelectric power (except for qualified hydropower as defined in the bill).	Total electricity sold, less hydroelectricity, less fossil fuel w/sequestration, less new nuclear and nuclear improvements.	Total # of kWhrs sold.
Target/goal	50% of base quantity by 2050.	50% of base quantity by 2050.	15% of base quantity by 2021.	15% of base quantity by 2020.
Interim targets/goals	Yes	Yes	Yes	Yes
Qualifying Energy Sources				
Biomass	Yes. Biomass that provides electrical and thermal output is incentivized to maximize efficiency. Biomass can receive up to 1.5 credits depending on efficiency.	Yes	Yes	Yes

	Clean Energy Standard Act of 2010 S. 20	Practical Energy and Climate Plan Act of 2010: Title III -Diverse Domestic Power S. 3464	Renewable Electricity Promotion Act of 2010 S. 3813	Substitute Amendment for H.R. 2454 (dated May 19, 2009)
Solar	Yes	Yes	Yes	Yes
Wind	Yes	Yes	Yes	Yes
Geothermal	Yes	Yes	Yes	Yes
Ocean	Yes	No	Yes	No
Landfill gas	Yes	Yes (also mentions "biogas")	Yes	Yes
Qualified hydropower	Yes. Defined as additional generation from efficiency improvements or capacity additions made on or after January 1, 1992. Capacity additions made on or after January 1, 2001. Small hydro (<50MW) in Alaska.	Yes. Additional energy from efficiency improvements or capacity additions, capacity additions to nonhydroelectric dams; new hydroelectric dams.	Identical to S. 20.	Yes
Marine & hydrokinetic	Yes	Yes	Yes	Yes

Table A-1. (Continued).

	Clean Energy Standard Act of 2010 S. 20	Practical Energy and Climate Plan Act of 2010: Title III -Diverse Domestic Power S. 3464	Renewable Electricity Promotion Act of 2010 S. 3813	Substitute Amendment for H.R. 2454 (dated May 19, 2009)
Incremental geothermal	Yes. Defined as the excess of total kWhrs produced from a geothermal facility over the average kWhrs produced at the facility for 5 of the 7 previous years (eliminate the highest and lowest kWhr production years).	No	Yes. Defined as the excess of total kWhrs produced from a geothermal facility over the average kWhrs produced at the facility for 5 of the 7 previous years (eliminate the highest and lowest kWhr production years).	No. Geothermal is listed as a clean energy source, but not in the context of 'incremental' geothermal as articulated in other proposals).
Coal-mine methane	Yes	Yes	Yes	Yes (specifically called "mine methane gas").
Qualified waste-toenergy	Yes. Defined as energy from combustion of post-recycled MSW or from the gasification or pyrolization of such waste and the combustion of the resulting gas at the facility.	Yes. Termed simply as "waste-toenergy."	Identical to S.20	No. Not specifically mentioned, but may qualify as a carbon-based fuel if 50% of CO_2 is captured and sequestered.

	Clean Energy Standard Act of 2010 S. 20	Practical Energy and Climate Plan Act of 2010: Title III -Diverse Domestic Power S. 3464	Renewable Electricity Promotion Act of 2010 S. 3813	Substitute Amendment for H.R. 2454 (dated May 19, 2009)
Qualified nuclear energy	Yes. Defined as a nuclear generating unit placed in to service on or after the date of enactment. Also includes "incremental nuclear" defined as additional generation from efficiency improvements or capacity additions.	Yes. Nuclear generating units placed in service after enactment of the proposed bill.	No	Yes. This proposal simply refers to "nuclear energy" and has no qualifications.
Eligible retired fossil fuel	Yes. Electricity generated by a fossil fuel generating facility with average CO_2 emissions >2,250 lbs per MWhr and is permanently retired between the enactment date and January 1, 2015.	Section 302, "Fossil Fuel Generating Facility Retirement Program," offers regulatory relief for early retirement of electric generating units. Program is managed by EPA.	No	No
Advanced Coal/ Fossil (w/CCS)	Yes. New or existing coal generation facility that permanently sequesters or stores at least 65% of GHGs.	Yes. Advanced Coal defined as a coal generating facility that captures, sequesters, stores, or reuses at least 80% of GHGs produced.	No	Carbon-based fuels qualify if at least 50% of carbon is captured, sequestered, or converted.

Table A-1. (Continued).

	Clean Energy Standard Act of 2010 S. 20	Practical Energy and Climate Plan Act of 2010: Title III -Diverse Domestic Power S. 3464	Renewable Electricity Promotion Act of 2010 S. 3813	Substitute Amendment for H.R. 2454 (dated May 19, 2009)
Combined heat and power	Yes. However, CHP counts towards electricity savings.	Not mentioned.	Yes. However CHP counts towards electricity savings.	Yes. However CHP counts towards "Energy Efficiency/savings" credits.
Re-powering or Cofiring increment	No. However, bill does include "incremental fossil fuel production" defined as additional generation from efficiency improvements or capacity additions that result in no additional GHG emissions.	No	No	Yes. Additional generation placed in service on or after January 1, 2001 to generate electricity from a clean energy source to include the portion of electricity generated from biomass co-firing.
Other Design Elements				

	Clean Energy Standard Act of 2010 S. 20	Practical Energy and Climate Plan Act of 2010: Title III -Diverse Domestic Power S. 3464	Renewable Electricity Promotion Act of 2010 S. 3813	Substitute Amendment for H.R. 2454 (dated May 19, 2009)
Energy efficiency/savings credits	Yes. Can be used to comply with 25% of clean energy targets.	Yes	Yes. Can be used to comply with 26.67% of requirement.	Yes. Can be used to comply with the entire goal/requirement if the governor of a state petitions the Secretary of Energy to allow energy efficiency/savings credits to be used for standard compliance.
Alternative compliance payments (ACP)	Yes. $0.035/kWhr annually adjusted for inflation.	Yes. Determined by the Secretary but not less than $0.05/kWhr + inflation adjustment. ACPs are paid directly to each State and funds may used for grants that increase the quantity of energy produced from diverse energy resources or offsetting costs to electricity consumers.	Yes. $0.021/kWhr (inflation adjusted). 75% of ACPs are paid to the State in which the utility is located and the governor of such State may use these funds to provide grants that increase the quantity of electricity generated from renewable sources, and/or to promote electric drive vehicles in the state, and to offset costs to electricity consumers.	No. Bill language mentions an alternative compliance mechanism, but in the context of payments made to a individual states and for compliance with State renewable portfolio standard programs.

Table A-1. (Continued).

	Clean Energy Standard Act of 2010 S. 20	Practical Energy and Climate Plan Act of 2010: Title III -Diverse Domestic Power S. 3464	Renewable Electricity Promotion Act of 2010 S. 3813	Substitute Amendment for H.R. 2454 (dated May 19, 2009)
Credit Trading	Yes. Clean Energy Credits and Energy efficiency credits.	Yes. However credits for end-user savings and energy efficiency savings cannot be sold outside of the state in which the electricity is generated.	Yes. Clean Energy Credits and Energy efficiency credits.	Yes
Double credits	Yes. (1) facilities on Indian land, (2) first 5 advanced coal facilities that sequester 1 million tons/yr of CO_2, (3) first 5 retrofitted coal plants that capture at least 200MWe of flue gas and sequester CO_2. If captured CO_2 is used for hydrocarbon recovery, then reduce credits by 0.25.	No	Yes. Projects on Indian land.	Yes. (1) Clean energy generation on Indian lands. Biomass co-fired with other fuels can receive double credits if biomass was grown on Indian land. (2) Distributed clean energy generation.

	Clean Energy Standard Act of 2010 S. 20	Practical Energy and Climate Plan Act of 2010: Title III -Diverse Domestic Power S. 3464	Renewable Electricity Promotion Act of 2010 S. 3813	Substitute Amendment for H.R. 2454 (dated May 19, 2009)
Triple credits	Yes. Small distributed generators <1MW on Indian land.	No	Yes. For small distributed generation on Indian land and projects that use algae biomass.	Yes. (1) Clean energy generation at an on-site facility that is used to offset part or all customer electricity requirements. (2) On-site eligible facility on Indian land can receive no more than 3 credits per kWhr.
Credits for advanced coal demonstration projects	Yes, by formula: Calculation = Total kWhrs supplied to grid multiplied by (CO_2 captured and sequestered/CO_2 captured & sequestered + CO_2 emitted).	Credits for demonstration coal projects during years 2015 – 2029. Projects must capture, permanently sequester, store or reuse 65% of greenhouse gases. Amount of credit calculated as: Total kWhrs to grid multiplied by CO_2 captured and sequestered/ (CO_2 captured & sequestered + CO_2 emitted).	No	No

Table A-1. (Continued).

	Clean Energy Standard Act of 2010 S. 20	Practical Energy and Climate Plan Act of 2010: Title III -Diverse Domestic Power S. 3464	Renewable Electricity Promotion Act of 2010 S. 3813	Substitute Amendment for H.R. 2454 (dated May 19, 2009)
Civil penalties (noncompliance)	# of kWhrs in violation multiplied by 200% of ACP (inflation adjusted).	Same as S.20.	Same as S.20.	Not specifically addressed in the proposed bill.
Exemptions	Utilities that sold less than 4 million MWhrs in the preceding calendar year. All utilities in Hawaii.	None	Same as S.20.	Retail electric suppliers that sold less than 1 million MWhrs in the preceding calendar year. All utilities in Hawaii. All federal, state, and municipal-owned utilities. All rural electric cooperatives.
Loans to support compliance	Yes	No	Yes	Yes

APPENDIX B. STATE-LEVEL BASELINE COMPLIANCE CALCULATION METHODOLOGY

Data Sources

Two EIA data sources were used to perform the baseline CES compliance assessment:

1) Electric Power Annual 2009—Data Tables: 1990-2009 Net Generation by State by Type of Producer by Energy Source (EPA 2009).[12]
2) 2009: EIA-923 January-December Final, Nonutility Energy Balance and Annual Environmental Information Data (EIA-923).[13]

Methodology

Calculating the generation mix for each state started with data from EPA 2009, which provides information regarding state-by-state electricity generation. A pivot table[14] was created to organize electricity generation data by state and by fuel source. However, the EPA 2009 data do not provide the detail necessary to distinguish between natural gas combined cycle (NGCC) generation and other natural gas generation technologies.[15] As a result, EIA-923 data were used to extract NGCC generation figures. With electricity generation from NGCC power plants now available, the pivot table from the EPA 2009 was modified to include NGCC generation and "Other Natural Gas" generation.

To be consistent with President Obama's Clean Energy Standard proposal, electricity generation sources were categorized as either "Clean Energy Generation" or "Other Generation."[16]

Energy sources categorized as "Clean Energy Generation" include:

*   Geothermal
*   Hydroelectric Conventional
*   Natural Gas Combined Cycle (50% of generation)[17]
*   Nuclear
*   Biomass
*   Pumped Storage

- Solar Thermal and Photovoltaic
- Wind
- Wood and Wood-derived fuels

Energy sources categorized as "Other Generation" include:

- Coal
- Natural Gas Combined Cycle (50% of generation)
- Natural Gas Other
- Other Gases
- Petroleum
- Other

In order to calculate the percent of generation from sources that qualify as "clean energy," the sum of "Clean Energy Generation" was divided by the total amount of generation. The same calculation was performed for "Other Generation." The pivot tables allowed this calculation to be done for the entire country as well as for each state.

End Notes

[1] A Renewable Energy Standard is a policy that requires electricity generation from "renewable" energy sources such as wind, solar, geothermal, biomass, etc. Many Clean Energy Standard proposals are broader and typically include renewable energy as well as other energy sources such as nuclear power and coal with carbon capture and sequestration.

[2] For specific information about a Federal Renewable Electricity Standard, see CRS Report R41493, *Options for a Federal Renewable Electricity Standard*, by Richard J. Campbell.

[3] Renewable energy typically includes biomass, solar, wind, and geothermal. The definition of "biomass" is somewhat different across the four proposals, with some proposals including a detailed description of what would be considered "biomass" and others simply referring to the definition provided in section 203(b) of the Energy Policy Act of 2005 (42 U.S.C. 15852(b)). For a more detailed discussion of biomass, see CRS Report R40529, *Biomass: Comparison of Definitions in Legislation Through the 111th Congress*, by Kelsi Bracmort and Ross W. Gorte.

[4] A megawatt is equal to 1,000 kilowatts.

[5] President Barack Obama, 2011State of the Union address, January 25, 2011, available at http://www.whitehouse.gov/the-press-office/2011/01/25/remarks-president-state-union-address.

[6] White House Office of Media Affairs, "President Obama's Plan to Win the Future by Producing More Electricity Through Clean Energy," February 3, 2011, available at http://www.whitehouse.gov/the-press-office/2011/02/03/president-obama-s-plan-win-future-making-american-businesses-more-energy.

[7] CRS was able to replicate this number by analyzing EIA's 923 electricity generation survey and Electric Power Annual—2009 data. Analysis results indicate that the 40% clean energy number announced by the President during his 2011 State of the Union address includes generation from the following energy sources: geothermal, hydroelectric, nuclear, biomass, pumped storage, solar thermal and photovoltaic, wind, wood/wood derived fuels. Also, a 50% CES credit per kilowatt hour was provided for natural gas combined cycle (NGCC) generation. The 50% credit for NGCC is based on emission data that indicates NGCC carbon emissions are about 50% less than coal carbon emissions. DOE's National Energy Technology Laboratory (NETL) reports the following: CO_2 emissions (lb/MWh $_{gross}$) for Supercritical Pulverized Coal = 1,675; CO_2 emissions (lb/MWh $_{gross}$) for NGCC = 790. See "Cost and Performance Baseline for Fossil Energy Plants Volume 1: Bituminous Coal and Natural Gas to Electricity," Department of Energy National Energy Technology Laboratory, November 2010, p. 5, available at http://www.netl.doe.gov/energy-analyses/pubs/ BitBase_FinRep_Rev2.pdf.

[8] Levelized cost of electricity (LCOE) is a methodology used to compare the cost of electricity generation from multiple energy sources while taking into account capacity factor, operations and maintenance, fuel cost, and financial cost differences. More detail along with DOE/EIA LCOE estimates for multiple energy sources can be found at http://www.eia.doe.gov/oiaf/aeo/electricity_generation.html.

[9] For detailed information about carbon capture technology, see CRS Report R41325, *Carbon Capture: A Technology Assessment*, by Peter Folger.

[10] The Center for American Progress has proposed that a federal Clean Energy Standard include a provision that requires 35% of U.S. total electricity generation come from "truly renewable" sources by 2035. See Richard W. Caperton, Kate Gordon, Bracken Hendricks, and Daniel J. Weiss, *Helping America Win the Clean Energy Race: Innovating to Meet the President's Goal of 80 Percent Clean Electricity by 2035*, Center for American Progress, February 7, 2011.

[11] Senators Jeff Bingaman and Lisa Murkowski, *White Paper on a Clean Energy Standard*, Committee on Energy and Natural Resources, United States Senate, March 21, 2011, available at http://energy CESWhitePaper.pdf.

[12] Data file available at http://www.eia.doe.gov/cneaf/electricity

[13] Data file available at http://www.eia.gov/cneaf/electricity

[14] A pivot table is a spreadsheet feature that organizes data in a spreadsheet database and groups data/information based on different parameters.

[15] President Obama's CES proposal specifically mentions "efficient natural gas" as a qualifying energy source. Natural Gas Combined Cycle (NGCC) is one of the most efficient natural gas electricity generation methods and this analysis assumes that "efficient natural gas" is synonymous with NGCC.

[16] "Other Generation" generally includes electricity generated from fossil energy.

[17] Analysis conducted assumes that Natural Gas Combined Cycle, since it is classified as "efficient natural gas," receives a 50% CES credit. While this 50% NGCC credit is not specifically described in President Obama's proposal (it does reference "partial credits for ... efficient natural gas.") Department of Energy NETL analysis indicates that NGCC generation emits 50% less carbon dioxide compared to coal (see footnote 7 above). This was the basis for the 50% partial credit for NGCC. NGCC generation consisted of electricity generation categorized in prime mover codes "CA", "CS", and "CT" in EIA Form 923.

In: Clean and Renewable Energy Standards   ISBN: 978-1-61324-932-1
Editors: B. J. Ruther, J. R. Moran   ©2012 Nova Science Publishers, Inc.

Chapter 2

OPTIONS FOR A FEDERAL RENEWABLE ELECTRICITY STANDARD[*]

Richard J. Campbell

SUMMARY

The choice of power generation technology in the United States is heavily influenced by the cost of fuel. Historically, the use of fossil fuels has provided some of the lowest prices for generating electricity. But growing concerns over greenhouse gas emissions and other environmental costs associated with burning fossil fuels are leading some utilities and energy providers to deploy more renewable energy technologies to meet power demands.

State governments have generally led the way in encouraging deployment of renewable energy technologies. Many states are essentially picking up where federal research and development dollars left off, using a Renewable Portfolio Standard (RPS) to create a market for renewable energy via mandatory requirements. While most RPS goals are expected to be met, about 12 states have existing provisions expiring by 2015, and approximately 14 states and the District of Columbia have existing RPS or related provisions scheduled to expire by 2020.

Wide-scale deployment of renewable energy technologies is at the heart of policy discussion for a national Renewable Electricity Standard

---

[*] This is an edited, reformatted and augmented version of a Congressional Research Service publication, CRS Report for Congress R41493, from www.crs.gov, dated November 12, 2010.

(RES), which would require certain retail electricity suppliers to provide a minimum percentage of the electricity they sell from renewable energy sources or energy efficiency. Green jobs growth from renewable and clean energy development is one of the goals of RES policy development; however, embedded energy efficiency requirements could also act to reduce the need for new renewable electricity generation facilities. An alternative Clean Energy Standard would provide incentives to certain advanced coal and nuclear facilities while also targeting retirement of older, polluting fossil fuel generation. Most of the opposition to an RES concerns the potentially higher cost to consumers of compliance using renewable electricity technologies.

The United States has traditionally relied primarily on market forces and temporary tax incentives to encourage the development and deployment of new technologies. This strategy is the "business as usual" model. However, several other forces are in play that call into question the "business as usual" model for innovation and deployment of renewable energy technologies. Even with generous tax incentives, non-hydro renewable electricity constituted approximately 4% of U.S. electric power industry capacity as of 2009. If renewable electricity is to play a larger role in the electricity future of the United States, many maintain that federal action may be necessary. As a result, some observers have argued for governmental intervention to bolster and accelerate U.S. activities relating to renewable energy.

A federal RES could offer an opportunity to drive renewable energy market growth by creating a compliance requirement nationally, bridging the gap of expiring or lower state RPS standards into future years. A Feed-in Tariff (FIT) is an alternative incentive concept to drive renewable energy growth via a mandatory purchase requirement by electric utilities. However, current U.S. law limits options for a national FIT.

The future global clean energy market has been estimated by 2020 to have sales as high as $2.3 trillion, and, as such, would be one of the world's biggest industries. Many nations are moving to secure a share of the expected rewards. Many argue what still appears to be missing is a longterm U.S. national energy policy that fully considers the costs and benefits of paths forward. The vision and clarity of a U.S. plan of action coming out of a well-defined national energy policy could provide the transparency and regulatory certainty the investment community has long claimed as necessary to help finance the modernization of the U.S. electricity sector.

INTRODUCTION

Electric power generation in the United States is currently dominated by the use of fossil fuels. Coal is the major fuel used to produce electricity from steam turbine-generators employing basic principles in use since the Industrial Revolution. However, burning coal results in environmental costs as emissions of nitrogen and sulfur oxides contribute to formation of smog and acid rain. Today, carbon dioxide emissions from the burning of coal and other fossil fuels are also widely believed to be contributing to global climate change and its potentially damaging effects.

Renewable energy has been used since long before the Industrial Revolution, but not on the scale of steam power generation. Renewable energy technologies use the power of the sun, wind, water, and heat from the earth, offering the possibility of producing electricity on a large scale without many of the environmental and climate consequences of electric power generation using fossil fuels. If harnessed by the right technologies, renewable energy offers the possibility for achieving inexpensive, almost limitless electricity with minimal adverse environmental impacts.

Much of the modern impetus for renewable energy development in the United States came from the Arab oil embargo of the 1970s, and the resulting energy crisis. A focus on national security and energy independence emerged with a goal of reducing dependence on foreign supplies of oil. While petroleum is not a major fuel source for electric power generation, the importance of energy to the national economy was underscored. Energy security and independence concerns, combined with growing reliability and environmental concerns led to the development of the renewable energy research and development programs at the U.S. Department of Energy (DOE). Today, added concerns over the global impacts of anthropogenic climate change and a desire for a lasting recovery from the recent recession have increased calls for a comprehensive national strategy making renewable energy a cornerstone of a policy for continuing U.S. economic development and jobs growth. Congress is currently considering legislation to address these concerns.

This report discusses current ideas for a federal Renewable Electricity (or Energy) Standard (RES) and a broader Clean Energy Standard (CES). The RES concept would require certain retail electric providers to obtain a minimum percentage of the power they sell from renewable energy sources or energy efficiency. The CES concept would extend the eligible technologies meeting the requirement to include advanced coal and nuclear energy, while also targeting retirement of existing polluting fossil generation. This report refers generally to both concepts as an RES unless otherwise stated. The goal of this report is to explore how such policies could potentially increase the amounts of renewable electricity generated in the United States, discussing other related public policy goals and rationales for renewable energy development, and the challenges/drawbacks of RES policy.

BACKGROUND

Electric Power Generation in the United States

The choice of power generation technology in the United States is heavily influenced by the cost of fuel. Historically, the use of fossil fuels has provided some of the lowest prices for generating electricity. Figure 1 shows that, as of 2009, coal accounts for approximately 45% of net generation by the electric power sector,[1] followed by natural gas at 23%, and nuclear power at 20%.

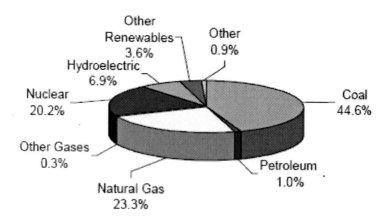

Source: U.S. Energy Information Administration, *Annual Energy Review* 2009, Table
    8.2a.
Notes: Total Net Generation is 3.953 billion kwh. See
    http://www.eia.gov/emeu/aer/elect.html.

Figure 1. U.S. Electric Power Industry Net Generation, 2009.

Electric power generation is responsible for 37% of U.S. domestic carbon dioxide emissions (the primary anthropogenic greenhouse gas (GHG)), and over one-third of all U.S. GHG emissions.[2] Growing concerns over GHG emissions, other environmental costs associated with burning fossil fuels, and existing or anticipated state and federal policies addressing these issues are leading some utilities and energy providers to deploy more renewable energy technologies to meet power demands. As of 2009, hydropower represented 7% of all U.S. electric power industry net generation with all other renewable energy accounting for a further combined total of 4%.[3]

Status and Potential of Renewable Energy in the United States

A summary of the status of current major renewable energy technologies follows, using figures for domestic estimated growth based on projections in the DOE's Energy Information Administration's (EIA's) *Annual Energy Outlook for 2010*.[4] Renewable energy technologies are at different stages of maturity in the technology development cycle, and are still being optimized from an engineering viewpoint. Some of the major perceived barriers the technologies must overcome to achieve a greater share of the electricity generation market are presented.

Biomass

Biomass for electric power is arguably the most conventional of all renewable electricity technologies and has a potentially large future. Biomass is currently the largest non-hydropower source of renewable energy consumed in the United States. Approximately 53% of all renewable energy used comes from biomass represented by biofuels, landfill gas, biogenic municipal solid waste, wood, wood-derived fuels, and other biomass such as switchgrass and poplar trees.[5] Agricultural wastes (such as corn stover) are another potential feedstock. With wood and biomass net summer capacity reported at 7 Gigawatts (GW) for 2009, DOE estimates that 29 GW of domestic biomass generation could be available by 2030.[6] Sustainable management of biomass resources, especially forests, will be critical to this future.

Biomass combustion is a relatively mature technology, but it is not widely used and is generally most efficient from a heat input-to-energy produced perspective when used in a combined heatand-power application.[7] Large scale co-firing of biomass with coal is a higher efficiency, lower per unit cost option.[8] Technologies for biomass gasification could potentially result in higher efficiencies when used to produce synthesis gas[9] or hydrogen for heat and/or power production. Demonstration and deployment of newer industrial gasification technologies is needed to scale-up plants and provide economical designs with high degrees of availability.[10] Huge potential exists for the biomass category, as it is generally regarded as a carbon-neutral source of energy.[11]

Wind

With about 10 GW of capacity installed, wind power was second only to natural gas in U.S. capacity additions in 2009.[12] The total installed and grid-connected wind power capacity as of 2009 in the United States reached approximately 35 GW.[13] DOE estimates that domestic wind power could reach a capacity of 68 GW by 2030 in its baseline scenario.[14]

Wind turbines are increasing in generating capacity, with domestic turbines in 2009 averaging 1.74 Megawatts (MW) in capacity.[15] When the wind is blowing at speeds which can be harnessed, electricity can be generated at prices nearly competitive with conventional fossil fuels. Since the wind doesn't blow all the time and varies in strength, average costs are higher and integration of large amounts of wind into an electricity grid has often been raised as an issue.[16] Backup generation in the form of natural gas combustion turbines has been the standby choice in some instances. Energy storage (using batteries or other means) is often suggested as a potential answer to deal with intermittency concerns. Since many of the best wind resource areas in the United States are far from population centers where the electricity will be used, the development of transmission facilities to carry power to population centers has been discussed as a prerequisite for wider development. Given these factors, DOE projects (with a deliberate and sustained national effort) that as much as 20% of the nation's electrical supply could be provided by wind energy by 2030. This would require wind power capacity to reach 300 GW, or a growth of over 280 GW over the next 21 years. Achieving such a prodigious goal would mean addressing significant challenges in technology, manufacturing, employment, transmission and grid integration, markets, and siting strategies.[17]

Offshore wind power in the United States is a fledgling industry, having just received federal authority in 2010 to go ahead with the first U.S. offshore wind farm in Nantucket Sound, off the Massachusetts coast. Known as the Cape Wind project, it is designed to operate 130 turbines from the German firm Siemens AG with a total capacity of 420 MW.[18] Permits for more than 2,476 MW capacity for offshore projects are pending as of 2009.[19] The overall potential for U.S. offshore wind power production capacity was estimated at 908 GW in 2005.[20]

Solar PV

Another renewable resource with enormous potential is sunlight. Sunlight is converted directly into electricity using solar photovoltaic (PV) cells which today are largely made from crystalline silicon. Research is underway to reduce the cost of PV cells using base materials other than silicon (such as cadmium telluride) and to improve manufacturing techniques which may increase the efficiency of solar cells.[21] Technologies to increase or concentrate the amount of sunlight in PV cells could raise the efficiency of the light-to-electricity conversion.[22]

Solar PV is widely used in a number of off-grid applications where distributed energy resources[23] are useful, and in peaking power applications to reduce power usage from the electric utility grid. Battery storage is important to off-grid usage to extend hours of usage past peak daylight. While solar PV installations only represented 1.37 GW of cumulative capacity in 2009, DOE estimates generating capacity from solar PV could reach almost 12 GW by 2030 in the United States.[24]

Since the amount of electricity that can be produced from solar PV generally depends on the intensity of sunshine (and the angle at which PV panels face the sun), the best potential exists for applications in sunny regions and lower latitudes. But the relative success of solar PV installations in Germany (with a total capacity of 8.9 GW in 2009)[25] prove the wider applicability of the technology in less-than-optimal climes.[26] Integration of PV cells and materials into building structures and designs could be a major step for the technology.

Concentrating Solar Thermal

Concentrating solar thermal technologies use mirrors to concentrate sunlight and generate heat usually for steam production. This steam is then used to generate electricity or provide high-temperature hot water for industrial or other process uses such as heating and cooling. Fairly large areas of land are needed, plus access to water since it is used for steam generation and cooling. Some novel applications heat air directly to generate thermal gradients which are harnessed to produce electricity. Solar thermal technologies are currently used for utility-scale power generation, but costs in the United States are considerably higher than prices for fossil-fuel power generation. Advances in the designs and materials of absorbers, reflectors and heat transfer fluids in next-generation solar thermal systems could potentially reduce costs to six cents per kilowatt-hour (kwh) by 2015.[27]

Solar power, like wind power, is considered a variable resource but solar power technologies can generate the most power when demand is highest— when the weather is hot and sunny. Concentrating solar power thermal plants with heat storage capacity are being increasingly considered for large central station generating plants in the sun-rich areas of the western United States, which could make such plants a base load[28] option.

Improved energy storage schemes would benefit both conventional power generation and off-grid applications in particular. New utility-scale solar thermal facilities will likely be located mostly in the southwestern region of the United States, an area where water resources may be under stress.[29] As such, EIA projects slow growth in domestic power generation from solar thermal technologies from 0.61 GW capacity in 2007 to 0.93 GW by 2030.[30]

Applications in residential and smaller industrial/commercial facilities are another area of potential solar thermal use. Solar hot water heaters are growing in use in the United States, but have seen much wider applications in parts of Europe and Asia. Additional research and development (R&D) investment may be needed to increase energy conversion efficiencies and bring down energy costs if smaller solar thermal systems are to become mainstream choices domestically.

Geothermal

Steam or hot water extracted from geothermal reservoirs in the Earth's crust can be used to generate electricity, or to provide thermal energy for heating or thermal processes. EIA projects this hydrothermal capacity could reach almost 4 GW by 2030, up from 2.44 GW in 2009.[31] Geothermal energy may no longer depend upon the availability of suitable naturally occurring geothermal resources. Enhanced Geothermal Systems (EGS) are man-made geothermal reservoirs. By drilling into the Earth's crust and injecting water to create steam, EGS offers the potential to produce geothermal energy almost anywhere, not just in areas where steam or hot water occur naturally, and offers the possibility for large scale generation of clean energy. Improvements in drilling technology can lower costs, and better fluid flow techniques can increase the amount of power generated.

Ground Source Heat Pumps (GSHP) are used mostly in residential applications and take advantage of the temperature difference between the ground and air. GSHP require a piping network to be buried underground to serve the customer's heating and cooling needs. Energy efficiency standards focused on home heating and cooling could lead to improved technologies and wider deployment of GSHP in new construction.

Hydroelectric Power

Only 2,400 of the 80,000 dams in the United States produce electricity.[32] Building a new hydroelectric power plant is expensive, and can face considerable opposition based on environmental concerns if a large dam is to be built.[33] As such, with most of the better sites already developed, DOE does not expect much growth in large conventional hydroelectric capacity. As of 2009, EIA reported conventional hydropower electric capacity at 77.2 GW.[34] However, DOE has identified approximately another 5,677 sites with the potential to generate about 30 GW of capacity using small and "low head" hydroelectric technologies.[35]

Other hydroelectric technologies are less mature. Opportunities to generate power with small elevation differences and low flow applications may need further R&D to optimize the hydroelectric generation potential. Hydrokinetic energy technologies generate power from the movement of water. Electricity can be generated from the flow of water in rivers, or additionally from the flow of released water at existing dams. Wave energy and tidal flow demonstrations of the various technologies to tap the power potential of coastal waters and estuaries are just beginning.

Renewable Electricity Costs Compared with Fossil Fuels

As described in the preceding sections, renewable energy technologies are designed to harness a variety of renewable energy resources with very different physical characteristics, such as the wind and the sun. Different technologies have seen different levels of investment over the years based upon various evaluations of potential and economic readiness to serve current markets or applications. The timeframe under consideration is important in any discussion of the potential for renewable energy technologies, for the technologies are at different stages in their development cycles and have attributes suited to different applications and locations. Consequently, the costs of generating electricity varies by type and maturity level of each renewable energy technology. The potential for deployment often depends upon local incentives and the quality of the renewable resource.

The cost of producing electricity from renewable energy sources is generally higher than electricity generated from fossil fuels when capacity factor[36] and operations and maintenance (O&M) costs are considered. The variable nature of some renewable energy sources generally results in lower capacity factors. However, the fuel component of O&M for most renewable energy sources is zero. Transmission costs vary for renewable energy and conventional generation technologies alike, depending on the distance from the generating power plant to where the power is consumed.

Estimated Levelized Cost of New Generation Resources, 2016.

Plant Type	Capacity Factor (%)	U.S. Average Levelized Costs (2008 $/megawatthour) for Plants Entering Service in 2016				
		Levelized Capital Cost	Fixed O&M	Variable O&M (including fuel)	Transmission Investment	Total System Levelized Cost
Conventional Coal	85	69.2	3.8	23.9	3.6	100.4
Advanced Coal	85	81.2	5.3	20.4	3.6	110.5
Advanced Coal with CCS	85	92.6	6.3	26.4	3.9	129.3
Natural Gas-fired						
Conventional Combined Cycle	87	22.9	1.7	54.9	3.6	83.1
Advanced Combined Cycle	87	22.4	1.8	51.7	3.6	79.3
Advanced CC with CCS	87	43.8	2.7	63.0	3.8	113.3
Conventional Combustion Turbine	30	41.1	4.7	82.9	10.8	139.5
Advanced Combustion Turbine	30	38.5	4.1	70.0	10.8	123.5
Advanced Nuclear	90	94.9	11.7	9.4	3.0	119.0
Wind	34.4	130.5	10.4	0.0	8.4	149.3
Wind – Offshore	39.3	159.9	23.8	0.0	7.4	191.1
Solar PV	21.7	376.8	6.4	0.0	13.0	396.1
Solar Thermal	31.2	224.4	21.8	0.0	10.4	256.6
Geothermal	90	88.0	22.9	0.0	4.8	115.7
Biomass	83	73.3	9.1	24.9	3.8	111.0
Hydro	51.4	103.7	3.5	7.1	5.7	119.9

Source: Energy Information Administration, Annual Energy Outlook 2010, December 2009, DOE/EIA-0383(2009)

Figure 2. Estimated Annual Levelized Costs of New Generation.

EIA recently estimated the average levelized cost[37] (in 2008 dollars) of power generation for new power plants entering service in 2016 for a variety of energy technologies.[38] As shown in **Figure 2**, a levelized cost per megawatt-hour (mwh) of $100.4 was estimated for conventional coal power generation, $83.1 for natural gas-fired conventional combined cycle, $119 for advanced nuclear, $119 for onshore wind, $191.1 for offshore wind, $396.1 for solar PV, $256.6 for solar thermal, $115.7 for geothermal, $111 for biomass, and $119.9 for hydropower. These estimates appear to show that some renewable energy technologies are competitive in cost and approaching parity with some fossil fuel power generation options. Others, however, are quite high-cost, suggesting the need for significant improvement in technology to become cost competitive.

RENEWABLE PORTFOLIO STANDARDS IN STATES

Many states are essentially picking up where federal research and development dollars left off, using a Renewable Portfolio Standard (RPS) to create a market for renewable energy via mandatory requirements or voluntary goals.

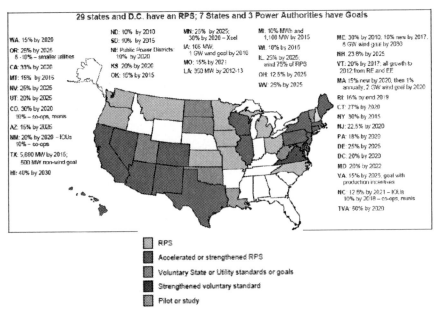

Source: Federal Energy Regulatory Commission, *Market Oversight*, August 2010.
Notes: Details including timelines may be found in the Database of State Incentives for Renewables and Energy Efficiency at http://www.dsireusa.org.

Figure 3. Renewable Power and Energy Efficiency Market: Renewable Portfolio Standards.

Through August 2010, RPS requirements or goals have been established in 29 states plus the District of Columbia.[39] RPS requirements generally oblige electric utilities to provide electricity from renewable energy sources in increasing amounts over a specified period of years. The requirement can be for a portion of the electricity provider's installed capacity, but most states specify a percentage of sales. Nineteen states and the District of Columbia have mandatory renewable energy requirements. Six states have renewable goals without financial penalties, and Virginia has incentives for utilities to meet renewable energy goals, and three power authorities have renewable energy goals. It should be noted that RPS and related policies in these states were designed to address various state objectives.[40]

EIA expects most states to meet existing RPS goals for renewable energy deployment.[41] But as is illustrated in Figure 3, about 12 states have existing provisions expiring by 2015, and approximately 14 states and the District of Columbia have existing RPS or related provisions scheduled to expire by 2020. The minimum amounts of renewable energy required in these RPS programs varies considerably, as does the timeframe for implementation and the eligible technologies.

DEBATE OVER A NATIONAL RENEWABLE ELECTRICITY STANDARD

The United States has traditionally relied on market forces and tax incentives to encourage the development and deployment of new technologies. This strategy is the "business as usual" model. However, several other forces are in play that call into question the "business as usual" model for innovation and deployment of renewable energy technologies. For example, investment dollars are relatively scarce at this time, as the nation struggles to emerge from a recession. Also, certain other nations, including Germany and China, have employed a different approach. They have aggressively used governmental policies to channel government resources into renewable energy programs that have permitted them to establish renewable energy industries whose products and productivity often exceed those of the United States.[42] The need for wide deployment of renewable energy technologies on a national scale is at the heart of policy discussion for a RES.

Advocates of a federal RES say it offers an opportunity to drive renewable energy market growth by creating a compliance requirement nationally, bridging the gap of expiring or lesser state RPS standards into future years. Legislation can impact energy markets by mandating change directly, i.e., through specific compliance requirements, or through incentives to encourage change. Such policy initiatives can be described as "market drivers" since they can cause changes affecting decisions regarding energy supply and demand. Many believe that the United States' existing (mostly non-renewable) low-cost energy system limits market opportunities in the short term, despite potential opportunities in the longer term both domestically and abroad. Some believe that energy policies based on temporary tax incentives are not true drivers of markets, since their effects may not last much past the expiration date of the incentive. Tax credits and other tax incentives can thus be seen as enablers of projects because they can add to a decision to go forward on a project. However, power projects still go forward without tax incentives if there is a market demand. Recent policies shifting certain renewable energy tax credits to cash grants have helped many new energy projects to be deployed, but how long these taxpayer-subsidized facilities will continue to operate absent continued support over the long term remains to be seen. Even with generous tax incentives and resource growth potential, non-hydro renewable electricity constituted approximately 4% of U.S. electric power industry production as of 2009. If renewable electricity is to play a larger role in the electricity future of the United States, many maintain that federal action may be necessary. As a result, some observers have argued for governmental action to bolster and accelerate U.S. activities relating to renewable energy. Renewable energy sources are seen as contributing to U.S. goals for energy security and energy independence by reducing dependence on petroleum imports with renewable electricity providing energy for electric vehicles. In addition, renewable biofuels can be used to generate electricity directly or power fuel cells used in electric or hydrogen-powered vehicles. Renewable energy sources can also improve the reliability of electric power systems by increasing the diversity of electricity generation resources,[43] and can potentially lower electricity prices by replacing more volatile fossil fuels.[44]

Opponents of a national RES base their arguments largely on cost. Electric power generation decisions in the past have largely been based upon the cost of generating electricity (excluding externalities—environmental costs to society not considered in energy prices). The cheapest fuel for many years has been coal, when economies of scale are considered. Coal currently accounts for almost 45% of net electricity generation in the United States, and replacing even a portion of that generation with renewable energy would not be a small undertaking. Renewable energy sources are naturally occurring but variable, and harnessing the forces of nature is not easy or very efficient given the current state of technologies. Most of the best large-scale hydropower resources have already been developed, and the best areas for wind power and solar energy are often far from population centers where the electricity produced would be consumed. Even if energy efficiency and energy conservation measures were to lessen the need for building future power generation, replacing large centralized power stations with renewable energy would be a massive task. Deployment of wind power (the fastest growing U.S. renewable energy segment) on a large scale would likely require considerable investments to upgrade electric transmission line capacity and install back-up power due to the intermittency of wind. Much of that back-up capacity would likely come from natural gas-fired combustion turbines, a fossil fuel with half the carbon intensity of coal but that comes with a history of price volatility.

Goals and Design of RES Policy

The key factors in the design of a RES policy include the final target level of the requirement (e.g., 5% or 25% of electricity sales), the time span over which the requirements are applicable, and the technologies being advanced by RES.

The level of the RES requirement can be a reflection of the rationale for the legislation, with deference to other considerations such as the relative quality of renewable energy resources in various regions of the United States. Recent RES proposals have been related to general goals to lower GHG emissions. The time span over which the RES proposal would be applicable reflects general goals for the legislation (for example, job creation or economic development).

Similarly, RES proposals have traditionally advanced renewable energy technologies which harness naturally replenished energy sources with a reduced or near zero GHG footprint. With GHG emissions reduction being a key feature of past RES proposal, recent legislative proposals have sought to expand the eligible technologies to include selected "clean energy" technologies such as mine-mouth methane or clean coal technologies.

ISSUES RELATED TO NATIONAL RES POLICY

The following paragraphs discuss both major and ancillary issues related to the implementation of a national RES policy on states, regions, and customers' income levels.

Regional Differences in Renewable Energy Resources

The quality (as defined by availability, magnitude and variability) of renewable energy resources differs in geographic regions across the United States. As a result, certain states and regions may be considered to be in a better position to comply with RES requirements at a lower cost than others, given the attributes of various renewable energy technologies.

Some electric utilities in the southeastern United States contend that renewable energy resources available in the region are insufficient to meet RES goals, potentially forcing the purchase of renewable energy from other regions with the result of higher prices for electricity. Forest-based biomass is seen as the most likely source for renewable energy development in the southeast, but competing uses for biomass raise questions regarding the sustainability of resource supplies.[45]

Similarly, concerns have been expressed in western states with good solar energy quality that large-scale development of solar thermal plants could stress constrained water resources.[46] These plants can consume as much water as comparable generating plants fueled by coal, if alternatives are not employed to reduce water used for steam production and cooling. Plans for developing solar thermal plants are therefore being viewed with water conservation in mind.

Governors in the eastern United States have voiced their own concerns regarding plans being discussed nationally for building high-voltage transmission lines to import wind power from the midwestern states.[47] The central issue is who will pay for the transmission lines.[48] Eastern states are wary of being assessed the entire cost of building such projects, and increased development of local renewable energy resources is seen by some as a better alternative.

RES and Infrastructure: Centralized vs. Distributed Electricity Generation

Compliance with an RES will likely require decisions on how goals for development of the electricity network should proceed to use renewable resources cost-effectively. RES requirements for energy efficiency could also lead to demand-side management[49] technology being incorporated into plans for infrastructure development. Such issues could impact how much new renewable generation will be required and where the generation will be located.

Most communities in the United States are served by a local electric utility which brings power to businesses and residences via electric transmission lines from steam-electric power generating facilities located miles away from large population centers. In fact, a network of power plants generally connects to a grid of higher voltage transmission lines which then distribute electric power over smaller lower voltage lines for customer use. Most of these plants burn coal or natural gas (i.e., fossil fuels) to generate electricity in base load operations. Efficiency of operation (the ratio of fuel input to energy produced) has been achieved from economies of scale, resulting in electrical power being generated fairly cheaply, but with a consequence of commensurately large GHG and other emissions. Nuclear power, though largely free of GHG emissions, is also drawn from large, centralized plants. This central station concept allows for a number of power plants to cost-effectively serve multiple communities, providing reliable power, and allowing different generating options to be incorporated into networks, including renewable electricity.

Some renewable electricity technologies are also capable of serving as base load capacity, notably geothermal, biomass, and hydroelectric dams. Wind power is considered intermittent because the wind doesn't blow all the time, and wind predictability is a concern especially for electricity dispatchers in organized markets (such as regional transmission organizations). Similarly, solar facilities depend on sunlight. Storage schemes may provide an answer for variability of power generation issues thus improving dispatchability, i.e., the capability to generate power to meet system loads.

Distributed generation schemes allow electric power to be generated at or near the point of consumption. Wind power and solar PV are well-suited to distributed generation, serving as backup or main power supplies for single customers or communities. Similarly, on-site or local generation may be seen as an option, and combined heat and power generation has long been used by industry to provide some or all of its electricity needs. Communities or consumers seeking back-up or independent power supplies using stand-alone generation may find this their best cost option. Renewable electricity technologies can take advantage of these opportunities.

The Smart Grid[50] is one of the options being discussed for the future of U.S. electricity networks and would build interactive intelligence into electricity transmission and distribution systems across the United States. Energy efficiency and energy conservation could be enhanced by demand-side management programs enabled by the wide scale deployment of smart meters.

Energy storage projects could enhance such a system, providing options for peak load management and potentially allowing for even greater cost savings. As such, aggressive deployment of a Smart Grid may even be able to reduce GHG emissions 18% by 2020.[51]

Transmission Infrastructure

The electricity transmission infrastructure of the United States is aging, and in many places is not keeping up with current uses, let alone future needs. Power interruptions and congestion problems cost the U.S. economy an estimated $100 billion a year in damages and lost business.[52] A study funded by the electric utility industry estimated that modernization of the grid may cost as much as $1.5 trillion to $2 trillion[53] if the electric system is to address projected future needs, potential national cyber security and other grid vulnerabilities,[54] and reliability concerns. Another recent study has looked at just the cost of upgrading the national transmission system to accommodate expected new renewable electricity generation through 2025.[55] The study concluded that incremental transmission to accommodate existing state RPS requirements will cost approximately $40 billion to $70 billion. To meet the higher target of either an existing state RPS or a 20% federal RES requirement, the study estimates incremental transmission costs of $80 billion to $130 billion.

Renewable energy projects in the western and southwestern United States are often located in areas far removed from large population centers where the power would be consumed. The process of selecting a route and siting a transmission line can take years even when only a single jurisdiction is involved. Building transmission lines across multiple states can be a very complicated and contentious process. The Energy Policy Act of 2005 established a process to identify and locate national interest energy transmission corridors (NIETCs) where transmission constraints or congestion may affect consumers.[56] The American Recovery and Reinvestment Act of 2009 (ARRA) later modified DOE's mission for NIETCs, directing DOE to include areas where renewable energy may be hampered by lack of access to the grid.[57] Some concepts for development of the Smart Grid could also allow for greater integration of intermittent or peaking renewable electricity generation. For example, advanced sensors built into an upgraded transmission system could monitor where such resources are generating electricity, and shift this power across the grid when and where it is needed.

The North American Electric Reliability Council has a mission to ensure reliability and is aware of the issues which lie ahead, describing the challenges of integrating renewable energy sources as requiring "significant changes to traditional methods for system planning and operation."[58]

RES and Options for Green Jobs Growth

Jobs growth from renewable and clean energy development is one of the greater goals of RES policy development.[59] An RES could provide a market driver for growth of renewable energy nationally, bridging the gap of expiring RPS standards and providing a driver for clean energy growth into future years. But modest RES requirements that slowly increase over the longer term may not provide a strong enough market driver in the short term to create domestic manufacturing opportunities on its own. Embedded energy efficiency requirements could also act to reduce the need for new renewable electricity generation facilities.

The domestic design and manufacture of renewable energy major equipment and components is the key to maximizing green jobs growth in the United States. Providing incentives to encourage manufacturers to locate production in the United States is therefore crucial to this intended future. Factories producing renewable energy equipment will likely require the development of supply chains, which in turn may spur expansion of a services sector to support the industry. Emergence of such supply chains and services could potentially result in a range of employment possibilities for both salaried and hourly wage workers.

National government investment on an enormous scale could be required to build a clean energy manufacturing future. Governments in Asia are investing amounts that are proportionally significantly greater than those spent by the United States [60] to develop clean energy manufacturing clusters.[61] Direct investments by governments in the Asia-Pacific region are aimed at replicating the success of the Silicon Valley cluster in Northern California where inventors, investors, manufacturers, suppliers, universities, and others established a "dense network of relationships." The goal of such clusters is to produce competitive, long-term cost and innovation advantages for participating firms and nations.[62] Supporters of a cluster development option believe that the United States cannot simply rely on an enhanced research, development, and deployment program to commercialize innovations needed to quickly expand its role in an international clean energy race. Many believe that ideally, these manufacturing clusters could be formed as public-private partnerships with state governments, businesses, and academic institutions as core members. One potential model could see federal funds competitively awarded to public-private partnerships seeking to establish such clusters around the United States, coordinated with such like-minded state government programs.

Longer-term, a focus of the Advanced Projects Research Agency-Energy[63] program is aimed at the next generation of clean energy technologies. Projects should aim for leap-frog advances in the renewable technologies developed if the goal is market share, for incremental improvements may not dislodge current market leaders. For example, if we assume a 25-year lifespan for today's utility scale wind power turbines and solar farms, this provides a target horizon for timing a roll-out of innovative clean energy technologies to replace the old technologies. The performance of these new technologies will need to be sufficiently "disruptive," so that decommissioning and replacement of the old technology will be seen as the most cost-effective decision.

The future global clean energy market has been estimated by 2020 to have sales as high as $2.3 trillion, and, as such, would be one of the world's biggest industries.[64] Many nations are moving to secure a share of the expected rewards. There are no guarantees that the commitment of resources by the United States to a clean energy manufacturing strategy will bring the desired results in future jobs or market share. Companies choosing to serve global markets have a variety of drivers in the decision about where to locate manufacturing facilities. Often, the potential size of the local market is a key consideration, and the countries which established (and those establishing) competitive clean energy industries began by serving a strong, often protected, domestic market. Trade policies will undoubtedly be a major factor in determining whether U.S. products can enter some of these global markets. Establishing incentives attractive enough for manufacturing to locate in the United States could help to create a strong export-focused industry.

Potential Cost Impacts on Lower Income Consumers

While specific strategies used by electricity retailers to comply with RES requirements would vary, a recent study suggests that an RES could result in dramatically higher costs to consumers.[65] The study asserts that if wind power and solar energy were to replace cheaper coal-fired electric generation, the additional costs of building transmission and replacement capacity to balance fluctuations when energy from intermittent renewable sources is unavailable could increase customer rates.

Current RES and CES proposals contain provisions for "Ratepayer Protection" which are intended to limit the incremental cost of compliance with an RES. However, any potential electricity rate increases would result in a relatively higher proportional impact on lower income customers. While both RES and CES proposals allow states the option of using "alternative compliance payments"[66] to offset potential increases in consumer bills caused by a RES, there is no provision directing states to use such funds specifically to relieve incremental cost impacts on lower income consumers.

National Feed-In Tariff Options

According to the National Renewable Energy Laboratory's report, "Feed-in Tariff Policy: Design, Implementation, and RPS Policy Interactions,"[67] feed-in tariffs (FITs) are the "most widely used policy in the world" to promote renewable energy deployment. FITs are an incentive policy to drive renewable energy growth via a mandatory purchase requirement by electric utilities, thus guaranteeing payments to renewable energy developers producing electricity. The NREL study suggests that an FIT can be a complementary strategy to an RES by focusing incentives on specific technologies.[68]

Feed-in tariff concepts are not new to the United States, as programs have been initiated by states such as Vermont, Oregon, and Washington, and by local municipalities such as Gainesville, FL, and Sacramento, CA. Many of these existing programs seek to provide incentives to a particular technology (most often solar PV), or seek to nurture domestic renewable energy industries.[69] A national FIT might seem unlikely for the United States quite simply due to the variability of renewable energy resources across the United States, and adoption of a FIT for one technology would inevitably favor specific regions.

While the Federal Energy Regulatory Commission (FERC) holds sway over interstate energy transactions and commerce, there are 50 state governments involved with local electricity issues and regional energy markets. FERC issued a ruling in July 2010[70] that sets the stage for how FIT concepts may be considered under current U.S. law. The California Public Utilities Commission (CPUC) wanted to implement a program under California state law establishing a FIT specifically for cogeneration systems under 20 MW. Several investor-owned utilities objected, maintaining that FERC alone had the authority to institute such a program. The CPUC asked FERC to clarify whether the Federal Power Act prevented the state from implementing the program. FERC concluded that the plan as proposed set wholesale rates in interstate commerce, which, as such, fell under federal jurisdiction. However, FERC went on to say that as long as cogenerators in this instance obtain "qualifying status" under the Public Utility Regulatory Policies Act of 1978 (PURPA , P.L. 96-617),[71] and the rates set by the CPUC do not exceed the "avoided cost"[72] of the utility purchasing the power, the proposal would not be preempted by PURPA or FERC regulations. California subsequently revised its FIT plans and proposed a new pilot program seeking to support 1 GW of renewable electric capacity. The plan requires the state's three investor-owned utilities to purchase power from renewable energy projects ranging from 1 MW to 20 MW through competitive auctions. Utilities are then required to award contracts with preference given to the lowest cost "viable" project, and then make subsequent awards to the next lowest cost project until MW requirements are reached.[73]

FERC's decision suggests that FIT policies can be applied to qualifying facilities (over 20 MW capacity) under PURPA , and still be subject to FERC and avoided cost pricing requirements. The price of renewable electricity appears to be falling to levels more competitive with conventional fossil generation. If FIT rates can be set at attractive price levels (within avoided cost requirements) allowing renewable energy developers to establish long-term contracts with electric utilities, then FITs may still be an option under current federal law.

Definition of a National Energy Policy

Given the major goals for an RES such as creating green jobs and reducing environmental impacts from traditional power generation using fossil fuels, enacting a clean energy strategy employing an RES or similar policy is an important option in the nation's energy strategy. Many argue what still appears to be missing is a long-term national energy policy which has fully considered the current and future energy needs of the United States, balanced by a deliberate evaluation of the costs (including externalities) and benefits (including employment). With recent technological developments raising the potential for natural gas to be produced from tight shale gas formations, the outlook for renewable electricity development could be affected if these unconventional natural gas sources can be developed and economically produced in an environmentally acceptable manner. The vision and clarity of a plan of action coming out of a well-defined U.S. national energy policy may provide the transparency and regulatory certainty the investment community has long claimed as necessary to help finance the modernization of the U.S. electricity sector.

INITIATIVES FOR A FEDERAL RENEWABLE ENERGY STANDARD

Several bills were introduced in the 111th Congress to establish a requirement for electric utilities to provide a fixed percentage of the electricity they sell to customers from renewable or clean energy sources. Two stand-alone proposals for a Renewable Electricity Standard (RES) and the similar proposal for a Clean Energy Standard (CES) are summarized in the following sections. CRS analysis of previous RES provisions in legislation pending before the 111th Congress is contained in the Appendix.

Summary and Analysis of S. 3813

S. 3813, the Renewable Electricity Promotion Act of 2010,[74] would amend PURPA,[75] adding a federal RES based on integrated renewable electricity and energy efficiency requirements.

Under S. 3813, each retail electric supplier with annual sales to electric consumers of 4 million megawatt-hours (mwh) or more would be required to earn or acquire renewable electricity or energy efficiency credits (RECs) for a portion of its annual retail sales. Electric utility companies in Hawaii would be exempt from the RES. Renewable energy from facilities on Indian Lands would be eligible to receive double RECs. Renewable energy from small distributed generation (less than 1 MW) and energy from algae would be eligible to receive triple RECs.

DOE would be required to establish a federal REC trading program to certify utility compliance with the RES. A REC could be traded, or held (carried forward) for up to three years from the date it was issued. One REC would be issued for each associated kwh of renewable energy or energy efficiency, and could be used only once for RES compliance. DOE could delegate its authority for the REC market function to a "market-making entity."

Renewable energy technologies qualifying under the RES included solar, wind, geothermal, ocean energy, biomass, landfill gas, qualified hydropower,[76] marine and hydrokinetic energy, incremental geothermal production, "coal-mined methane", qualified waste-to-energy, or other innovative renewable energy sources.

S. 3813 excludes from the base qualification calculation (i.e., the amount of annual electricity supply that the renewable energy and energy efficiency percentages would be applied to) the annual retail electricity sales power generation from hydroelectric facilities, nuclear power, and fossil electric power generation proportional to its GHG emissions captured and geologically sequestered.[77]

The minimum required total annual renewable electricity and energy efficiency percentages for the specified calendar year are:

2012 – 2013	3%
2014 – 2016	6%
2017 – 2018	9%
2019 – 2020	12%
2021 – 2039	15%

Energy efficiency would be limited to account for no more than 26.67% of the annual compliance requirement. Qualifying energy efficiency improvements would include customer facility energy savings, distribution system electricity savings, and incremental electric output from new combined heat and power systems compared to output from separate electric and thermal components. DOE would be required to issue regulations for measurement and verification of electricity savings. Compliance with other energy efficiency standards (whether federal, state or local) would not count toward the RES requirements.

An "alternative compliance payment" (ACP) of 2.1 cents per kwh (adjusted for inflation) would be allowed to meet RES requirements, with 75% of ACP being paid into individual state funds for a renewable energy escrow account. These funds would be designed to help states develop new renewable energy resources, energy efficiency, or promote the development, deployment, and use of electric vehicles and their batteries. States may also use the funds to offset increases in customer bills caused by the RES.

A civil penalty for non-compliance with RES requirements would apply, and would be calculated as the product of the number of kwhs sold to electric consumers in violation of requirements, multiplied by 200% of the ACP.

Utilities may petition DOE annually to waive compliance with RES requirements (in whole or part) for reasons of "Rate Payer Protection" to limit the rate impact of the incremental cost of compliance "to not more than 4% per retail customer" in any year. States may also petition DOE to seek a variance from compliance for "one or more years" due to "transmission constraints preventing delivery of service."

States may require higher renewable energy or energy efficiency levels, but all states must comply with RES requirements at a minimum. DOE would be required to facilitate cooperation between federal and state renewable energy and energy efficiency programs to the maximum extent practicable.

DOE may make loans to electric utilities for qualifying projects to facilitate RES compliance, or reduce the impact of RES requirements on customer electricity rates.

Beginning in 2017, and every five years thereafter, DOE would review and report to Congress on whether the program established by S. 3813 contributes to an "economically harmful" increase in electric rates in regions of the United States, analyze whether the program has resulted in "net economic benefits for the United States," and analyze whether new technologies and clean renewable sources would "advance the purposes of the section." DOE is to make recommendations on whether the percentages of energy efficiency and renewable electricity required should be increased or decreased, and whether the definition of renewable energy should be expanded to reflect changes in technology or whether previously unavailable resources of clean or renewable electricity should be increased or decreased.

The definition of biomass would be modified and renewable energy would be defined for purposes of the federal renewable energy purchase requirement. Sustainability would also be added as a focus for biomass harvesting, with a mandatory interagency (i.e., Department of Agriculture, Department of the Interior, and Environmental Protection Agency) report to Congress that assesses the impacts of biomass harvesting for energy production.

Comments on the S. 3813 RES

For the base qualification calculation of annual retail electricity sales power generation, S. 3813 would exclude fossil electric power generation in proportion to the GHG emissions captured and geologically sequestered. However, it is possible that captured GHG emissions may find other market-worthy applications if, for example, captured carbon dioxide can be economically converted to fuels or other useful products.[78] The bill does not address whether captured, non-sequestered but usefully applied GHGs may also be eligible for the same exemption.

Many municipal utilities and electric cooperatives fall below the threshold of 4 million mwh electricity sales. These entities would not have to comply with the RES.

Energy from algae would be eligible to receive triple RECs. The RES is focused on producing electrical energy. While most current algal biomass research seems to be leading to biofuels development, the opportunity to make a biofuel for producing electricity may exist (for example, to produce a "synthetic" natural gas or fuel for combustion turbines).

Transmission access could be a key issue in future renewable energy development. Under S. 3813, states may also petition DOE to seek a variance from compliance for "one or more years" due to "transmission constraints preventing delivery of service." There is no mention of how such a transmission constraint would be determined. For example, it does not specify whether the FERC would be involved in making the determination.

"Clean" energy is mentioned in the title of the legislation and with regard to future technologies eligible for the RES. Clean energy is differentiated from renewable energy in the bill as clean or renewable energy, but is not defined in the legislation. Nuclear and advanced coal technology projects are eligible as beneficiaries of ACP funds set up by states.

S. 1462, the American Clean Energy Leadership Act of 2009 (ACELA), proposed an RES provision which served as the model for the provisions in S. 3813. Differences between the two proposals include:

- The base quantity of electricity in ACELA would exclude electricity from incinerated municipal solid waste (MSW) owned by an electric utility or sold to an electric utility under a contract or rate order to meet the needs of its retail customers. In S. 3813, MSW qualifies only as a renewable energy resource under "qualified waste-to-energy."[79]
- ACELA returns all of the ACP to the state in which the electric utility is located, as opposed to S. 3813 in which 75% of the ACP is returned. S. 3813 does not specify whether amounts from civil penalties for non-compliance also go to state ACP funds.
- The start date for RES minimum targets for annual renewable electricity or energy efficiency in S. 3813 would be 2012 instead of

2011. Required annual percentages of sale to be met by renewable energy or energy efficiency are the same.

S. 20 Compared with S. 3813[80]

The stated goal of the CES in the Clean Energy Standard Act of 2010[81] in S. 20 is to support and expand the use of clean energy and energy efficiency, reduce GHG emissions, and reduce dependence on foreign oil. The bill would amend PURPA by adding requirements for a federal CES.

In S. 20, the base qualification of retail sales to electricity consumers excludes power produced from existing electric utility-owned hydroelectric facilities, and power generated from incineration of municipal solid waste facilities owned by electric utilities.

Clean energy has a broad definition in S. 20. In addition to the renewable energy technologies qualifying in S. 3813, several additional "clean energy" technologies would qualify for the CES including qualified nuclear energy, advanced coal generation, eligible retired fossil fuel generation, or other innovative clean energy sources as determined by the DOE Secretary in rulemaking.[82]

Clean energy credits in S. 20 would be issued for compliance with CES requirements, and could be banked for use in any future year or traded in a trading program to be established by DOE, with generally one credit being issued per kwh of associated clean energy generation or energy efficiency.

In S. 20, power from "eligible retired fossil fuel generation" is included in the clean energy definition as a mechanism to encourage the retirement of these facilities. Eligible fossil fuel power generation can be derived from any fossil fuel. The bill considers the quantity of electricity generated in the three years prior to retirement with average carbon dioxide emissions in excess of 2,250 pounds per mwh. Once a plan is in place to permanently retire the power plants by 2015, these facilities would be eligible for clean energy credits. Such credits would be issued at a rate of 0.25 credits per kwh for the three-year period beginning on the date of retirement of the facility, and the credit calculation would be based on the average annual quantity of electricity generated during in the three-year period prior to retirement.

Advanced coal generation in S. 20 would be eligible for clean energy credits based on kilowatt-hours net generation exported to the grid in the prior year multiplied by the ratio of carbon dioxide captured and sequestered compared to the total carbon dioxide captured, sequestered, and emitted.

Double credits would go to the first five advanced coal plants geologically sequestering at least one million tons per year of carbon dioxide. Coal plants "retrofitted" with advanced coal technology could also receive double credits for geologically sequestering the flue gas emissions equivalent to 200 MW of electric power generation. Carbon dioxide captured and used for enhanced oil recovery would receive credits reduced by a factor of 25%.

The definition of eligible biomass in S. 20 differs from S. 3813, which uses the definition from §203(b)(1) of the Energy Policy Act of 2005. The CES in S. 20 more closely follows the definition in the Food, Conservation, and Energy Act of 2008[83] and would include biomass removed from the National Forest System and other public lands under specific conditions generally considered to aid healthy forests or reduce the risk of forest fires.[84]

Combined heat and power facilities are rewarded in S. 20 with additional clean energy credits for higher efficiencies. One additional credit per kwh would be provided for systems that exceed a 50% efficiency improvement. Systems achieving a 90% improvement would qualify for 1.5 credits per kwh.

The minimum required annual clean energy or energy efficiency percentages in S. 20 for the specified calendar year are:

2013 – 2014	13%
2015 – 2019	15%
2020 – 2024	20%
2025 – 2029	25%
2030 – 2034	30%
2035 – 2039	35%
2040 – 2044	40%
2045 – 2049	45%
2050	50%

Energy efficiency would be allowed to count for no more than 25% of the annual compliance requirement in the CES. Potential electricity savings would be expanded to include savings from incremental nuclear and incremental fossil fuel production.

Alternative compliance payments in S. 20 are allowed to meet CES requirements at a rate of 3.5 cents per kwh, with all of the payment going to the state or states in which the electric utility operates in proportion to the base quantity of retail electricity in each state.

There is no amendment to federal purchase requirements in the CES.

Other Comments on the CES

The CES in S. 20 is more "aggressive" than the RES in S. 3813 in terms of applicable technologies (extending the program to cover certain base load electric generation technologies), with higher CES annual target percentage requirements (rising to 50% of an electric utility's annual retail sales), and overall length of the period of compliance (to 2050). Embedded is a strategy to address retiring older, inefficient fossil energy plants.

The desired balance in the CES between renewable electricity deployment and deployment of other clean energy technologies is not specified as is, for example, the divide between renewable electricity and energy efficiency projects.

It is not clear how the goal for advanced coal retrofits (i.e., 200 MW of "equivalent" flue gas emissions) considers the power production capacity of the plant prior to the retrofit.

APPENDIX. COMPARISONS OF PENDING CLEAN ENERGY INCENTIVE PROPOSALS

S. 1462 (American Clean Energy Leadership Act of 2009) and H.R. 2454 (American Clean Energy and Security Act of 2009)[85]

Sec. 132 of S. 1462 would establish a federal renewable electricity standard (RES) for electric utilities that sell electricity to consumers (for purposes other than resale). Such utilities must obtain a percentage of their annual electricity supply from renewable energy sources or energy efficiency, starting at 3% in 2011 and rising incrementally to 15% by 2021. Eligible renewable sources are defined as wind, solar, geothermal, and ocean energy; biomass, landfill gas, qualified hydropower (i.e., incremental additions since 1992), marine and hydrokinetic energy, coal-bed methane, and qualified waste-to-energy. Other types of renewable energy resulting from innovative technologies may be qualified by the Secretary of Energy via a rulemaking.

Under S. 1462, RES requirements are to be met by the annual submission of federal renewable energy credits (RECs), but up to 26.67% of the requirement may be met by energy-efficiency credits (EECs) in any one year (following a petition by a state's governor). Alternative compliance payments (ACPs) of 2.1 cents per kilowatt-hour are permitted in lieu of meeting the renewable electricity standard, with these payments going directly to the state in which the electric utility is located. Trading of RECs is permitted, and banking of RECs is allowed for up to three years; RECs are retired when submitted for compliance. EECs are awarded for electricity savings verifiably achieved by the electric utility's actions. The Secretary of Energy will provide guidelines and regulations for measurements and baseline definitions in the award of EECs. No EECs will be awarded for compliance with conservation or energy-efficiency standard programs.

Comparison to Similar Provisions in H.R. 2454,
American Clean Energy and Security Act of 2009

The structure and definitions of the Renewable Electricity and Energy Efficiency provisions in H.R. 2454 and S. 1462 are essentially the same with regard to eligible renewable energy technologies. Incremental hydropower added after 1992 can be considered renewable energy under the Senate version, as opposed to after 1988 in the House version.

S. 1462 requires compliance with its renewable electricity standard to begin in 2011, one year earlier than the House version. The State of Hawaii is exempted from compliance in the Senate bill. The Senate requirement advances to a maximum of 15% renewable electricity (of which energy efficiency may constitute as much as 26.67%); the House requirement has a maximum of 20% renewable electricity, of which up to 25% may come from energy efficiency.

The implementing agency is designated as DOE in the Senate bill, while the House version has FERC implementing the provision. Retail electric suppliers may receive RECs for complying with a state RES by generating or buying renewable electricity under the Senate bill, but not in the House bill. The Senate energy bill has no parallel provision to the House bill's recognition of renewable energy programs implemented by states which centrally purchase renewable energy.

The alternative compliance payment is 2.1 cents per kilowatt-hour (kwh) in the Senate Energy bill, compared with 2.5 cents per kwh in the House version. ACP funds can be used for nonrenewable energy deployment or energy efficiency under the Senate Energy bill, with generation from nuclear, coal with carbon sequestration and storage, and electric vehicle deployment being eligible. Direct grants to customers to offset higher costs from the RES are also allowed by the Senate bill from ACP funds. The House does not allow for a waiver of RES requirements, while the Senate energy bill allows for deferral due to extremes of weather or nature, to avoid utility rate incremental impacts of more than 4% in any year, or because of transmission constraints preventing delivery of service. There is no provision in the House bill for loans to help electric utilities comply with the RES.

The House bill increases the federal renewable energy purchase requirement beginning in 2012 to 6%, raising it to 20% by 2020, where it would remain through 2039. The Senate energy bill version stays with the lesser requirements in the Energy Policy Act 2005.

The House bill defines one renewable energy credit as representing one megawatt-hour of renewable electricity; a similar definition appears to be implicit (but is not specified) in the Senate energy version. Both renewable energy and energy-efficiency credits can be traded under the Senate bill, while only renewable electricity credits can be traded under the House legislation.

Triple credits are granted when electricity is provided through distributed generation (DG). Definitions of distributed generation eligible for triple RECs differ between the two bills. The Senate energy bill defines DG systems as being at or near a customer site, providing electric energy to one or more customers for purposes other than resale to a utility through a net metering arrangement. The House version defines DG as a facility that generates renewable electricity, primarily serving one or more electric consumers at or near the facility site, which is no larger than 2 MW at the time of enactment (or 4 MW after enactment), generating electricity without combustion. This rules out biomass or municipal solid waste combustion as eligible sources of DG. Both provisions require electricity generation, thus ruling out thermal applications (for example, hot water or steam systems). While not specifying a size limit on DG systems, the Senate only gives triple RECs to DG systems smaller than 1 MW, while the House gives triple RECs to all eligible DG systems.

The two bills differ in the exclusions that would be allowed from the calculation of a utility's total annual electricity supply, called the "base quantity of electricity." This is the amount of annual electricity supply that the renewable energy and efficiency percentages would be applied to. By reducing the annual base quantity, the exclusions would also reduce the total amount of renewable energy and efficiency that would be required. Both bills exclude existing hydro (except qualified hydro), nuclear capacity placed in service after the date of enactment, and the quantity of electricity in a CCS facility proportional to the amount of greenhouse gases (GHGs) sequestered. The Senate energy bill additionally excludes capacity of a municipal solid waste facility owned by, or sold under contract/rate order to, an electric utility, and nuclear power plant efficiency improvements and capacity additions made after the date of enactment.

S. 1462 (American Clean Energy Leadership Act of 2009) and S. 3464 (Practical Energy and Climate Plan Act of 2010)

S. 3464 would create a federal Diverse Energy Standard (DES) for electric utilities selling power to end-use customers. Utilities must obtain minimum annual percentages of the electricity they sell from energy efficiency, renewable energy or other [clean] energy sources of:

2015 – 2019	15%
2020 – 2024	20%
2025 – 2029	25%
2030 – 2049	30%
2050	50%

These diverse energy sources can include advanced coal generation, biomass, coal mine methane, end-user energy efficiency, efficiency savings in power generation, geothermal energy, landfill and biogas, marine and hydrokinetic energy, qualified hydropower (incremental capacity or efficiency improvements made up to three years prior to enactment), qualified nuclear (placed in service on or after date of enactment), solar, waste-to-energy, wind, and any other energy source that results in at least an 80% reduction in greenhouse gas emissions compared to average emissions in the prior year from "freely emitting sources."

S. 1462 would establish a federal Renewable Electricity Standard for electric utilities selling power to end-use customers. These utilities must obtain an annual percentage of their supplies from renewable energy sources or energy efficiency ranging from 3% in 2011 to 15% by 2021. Renewable sources are defined as wind, solar, geothermal, and ocean energy; biomass; landfill gas; qualified hydropower (i.e., incremental additions since 1992); marine and hydrokinetic energy; coal-bed methane; and qualified waste-to-energy.

Federal Clean Energy Standards
- S. 1462 would establish an RES for electric utilities selling power to end-use customers, requiring energy efficiency measures or renewable energy sources to start at 3% in 2011, rising to 15% of all resources by 2021.

- S. 3464 would create a DES for electric utilities selling electricity to end-use customers, requiring energy efficiency or clean energy sources to start at 15% by 2015, rising to 50% of all sources by 2050.

Other Renewable or Clean Energy Provisions

- S. 1462 excludes from the base quantity (to which RES requirements apply) electricity generated by electric utility-owned hydropower, incineration of municipal solid waste, and electricity from fossil fuel units proportional to greenhouse gas emissions captured and geologically sequestered. S. 3464 only excludes hydropower from the base quantity for the DES.
- S. 1462 would modify the requirement established in the Energy Policy Act of 2005 that federal agencies purchase and/or produce and use renewable electricity. The bill also promotes renewable energy development on federal lands and requires the establishment of Renewable Energy Permit Coordination Offices in field offices of the Bureau of Land Management in a pilot project to coordinate federal permits for renewable energy and electricity transmission.
- Only incremental hydropower, efficiency improvements or powering of non-hydroelectric dams is allowed for the definition of "qualified hydropower" in S. 1462. The definition in S. 3464 also allows for new hydroelectric dams to be included.
- Alternative compliance payments for the DES would be set at a minimum 5 cents per kilowatt-hour, which would be higher than the 2.1 cents per kwh set in S. 1462.
- Neither S. 1462 nor S. 3464 clearly defines the basis for issuance or award of a federal renewable energy credit or a diverse energy credit (DEC). While a REC appears to be issued for each megawatt-hour of renewable electricity in S. 1462, both bills associate such credits with a kwh of electricity "used only once" for compliance purposes. DES alternative compliance payments also associate each DEC with a "megawatt hour of demonstrated total annual electricity savings."

End Notes

[1] Includes electric utilities and independent power producers.
[2] U.S. Energy Information Administration, *Emissions of Greenhouse Gases in the United States 2008*, DOE/EIA-0573(2008), December 2009, http://www.eia.doe.gov/oiaf/1605/ggrpt/pdf/0573(2008).pdf.

[3] U.S. Energy Information Administration, *Electricity Net Generation: Total (All Sectors), 1949-2009*, Annual Energy Review 2009, Report No. DOE/EIA-0384(2009), July 19, 2010, http://www.eia.gov/emeu/aer/elect.html.

[4] U.S. Energy Information Administration, *Annual Energy Outlook for 2010 - Yearly Projections to 2035*, Renewable Energy Generating Capacity, May 2010, http://www.eia.doe.gov/oiaf/forecasting. Hereafter referred to as AEO 2010.

[5] CRS Report R41440, *Biomass Feedstocks for Biopower: Background and Selected Issues*, by Kelsi Bracmort.

[6] AEO 2010.

[7] Combined heat and power (also called "cogeneration") involves the production of electricity and use of rejected heat energy for a thermal application from a single use of fuel.

[8] Larry Eisenstat, Andrew Weinstein, and Steven Wellner, *Biomass Cofiring: Another Way to Clean Your Coal*, Power Magazine.com, July 1, 2009, http://www.powermag.com/coal/

[9] Synthesis gas (or "syngas") is a mixture of mostly carbon monoxide and hydrogen, and is chemically similar to natural gas.

[10] Availability represents the number of hours a power production facility is able to produce electricity relative to the total number of hours in the time period considered. Base load power plants fueled by coal or geothermal or nuclear energy may have an availability of between 70% and 90%.

[11] Biomass combustion is generally considered carbon neutral because trees and plants are considered to take in as much or more carbon dioxide in the growing cycle as they release when burned. However, an issue often raised is the timing of the release of carbon dioxide. The growing cycle takes in carbon dioxide as the plant grows, while the combustion process releases the stored carbon dioxide all at once.

[12] Ryan Wiser and Mark Bolinger, *2009 Wind Technologies Market Report*, Lawrence Berkeley National Laboratory, August 2010, http://eetd.lbl.gov/ea/ems/reports/lbnl-3716e-ppt.pdf.

[13] Global Wind Energy Council, *Global Wind 2009 Report*, March 2010, http://www.gwec.net/fileadmin/documents/Publications/Global_Wind_2007_report/GWEC_Global_Wind_2009_Report_LOWRES_15th.%20Apr..pdf.

[14] AEO 2010. Op. cit.

[15] Wiser and Bolinger, Op. cit.

[16] International Energy Agency, *Variability of Wind Power and Other Renewables: Management Options and Strategies*, Management Options and Strategies, 2005, http://www.iea.org/Textbase/Papers/2005/variability.pdf.

[17] U.S. Department of Energy, Office of Energy Efficiency and Renewable Energy, *20% Wind Power by 2030: Increasing Wind Energy's Contribution to U.S. Electricity Supply*, December 2008, http://www1.eere.energy.gov/windandhydro/pdfs/42864.pdf.

[18] Reuters, "Cape Wind, First U.S. Offshore Wind Farm, Approved," April 28, 2010, http://www.reuters.com/article/ idUSTRE63R42X20100428.

[19] See http://eetd.lbl.gov/ea/ems/reports/bnl-3716e-es.pdf.

[20] Walt Musial, *Offshore Wind Energy Potential for the United States*, National Renewable Energy Laboratory, May 2005 http://www.windpoweringamerica.gov/pdfs/workshops/2005_summit/musial.pdf.

[21] Solar Today, *Game Changing Technology on the Horizon*, American Solar Energy Society, March 2009, http://www.solartoday-digital.org/solartoday/200903/?pg=27.

[22] Andrew Moseman, *How Solar Power Could Become Cheaper Than Coal*, Discover magazine, November 1, 2010, http://discovermagazine.com/2010/jul-aug/01-how-solar-power-become-cheaper-than-coal

[23] [Distributed generation] is located close to the particular load that it is intended to serve. General, but non-exclusive, characteristics of these generators include an operating strategy that supports the served load and interconnection to a distribution or sub-transmission system (138 kV or less). See http://www.eia.doe.gov/glossary/index.cfm?id=D.

[24] AEO 2010. Op. cit.

[25] Sustainable Business.com News, *US Solar Capacity Grew 37% in 2009*, April 16, 2010, http://www.sustainablebusiness.com/index.cfm/go/news.display/id/20148.

[26] Australian Broadcasting Corporation, "Germany an Unlikely Hot Spot for Solar Power," August 2007, http://www.abc.net.au/news/stories/2007/08/01/1994041.htm.

[27] NREL 2009. Op. cit.

[28] A base load plant, usually housing high-efficiency steam-electric units, normally operates to take all or part of the minimum load of a system, and consequently produces electricity at an essentially constant rate and runs continuously. These units are operated to maximize system mechanical and thermal efficiency and minimize system operating costs. See http://www.eia.doe.gov/glossary/index.cfm?id=B.

[29] Solar thermal power plants currently use almost as much water as fossil-fueled power plants. See CRS Report R40631, *Water Issues of Concentrating Solar Power (CSP) Electricity in the U.S. Southwest*, by Nicole T. Carter and Richard J. Campbell.

[30] AEO 2010. Op. cit.

[31] AEO 2010. Op. cit.

[32] Oak Ridge National Laboratory, *Dams: Multiple Uses and Types*, http://www.ornl.gov/info/ornlreview/rev26-34/text/hydside1.html.

[33] CRS Report R41089, *Small Hydro and Low-Head Hydro Power Technologies and Prospects*, by Richard J. Campbell.

[34] AEO 2010. Op. cit.

[35] Idaho National Laboratory, Feasibility Assessment of the Water Energy Resources of the United States for New Low Power and Small Hydro Classes of Hydroelectric Plants, January 2006, http://hydropower.inel.gov/resourceassessment/ pdfs/main_report_appendix_a_final. pdf.

[36] Capacity factor is the ratio of the electrical energy produced by a generating unit for the period of time considered to the electrical energy that could have been produced at continuous full power operation during the same period. See http://www.eia.doe.gov/glossary/index.cfm?id=C.

[37] The present value of the total cost of building and operating a facility over its life, as represented by equal annualized amounts.

[38] U.S. Energy Information Administration, *2016 Levelized Cost of New Generation Resources from the Annual Energy Outlook 2010*, AEO 2010, January 12, 2010, http://www.eia.doe.gov/oiaf/aeo/electricity

[39] Federal Energy Regulatory Commission, *Renewable Power & Energy Efficiency Market: Renewable Portfolio Standards*, Renewable Portfolio Standards (RPS) and Goals, August 2010, http://www.ferc.gov/market-oversight/

[40] There can be multiple goals for an RPS, and some states aim for a broader set of objectives than others. Examples of broader goals and objectives include local, regional, or global environmental benefits; local economic development goals; hedging fossil fuel price risks; and advancing specific technologies. See http://www.epa.gov/chp/state renewable_fs.html.

[41] Energy Information Administration, *Annual Energy Outlook 2010 with Projections to 2035*, Report #:DOE/EIA-0383(2010), May 11, 2010, http://www.eia.doe.gov/oiaf/aeo/state

[42] BusinessGreen.com, "UK and US lag China and Germany in Race to Attract Clean Tech Investors," October 28, 2009, http://www.businessgreen.com/articles/print/2252151.

[43] The Federal Energy Regulatory Commission (FERC) accepted a California Independent System Operator proposal that would allow "intermittent" generation resources to participate in the state's competitive spot market. Intermittent resources are solar or wind power (or other power-production technologies) systems that produce electricity only under certain conditions or during certain hours. FERC said in its decision under docket numbers ER02-922-000 and EL02- 51-000, "Encouraging the development of intermittent generation will increase diversity in the resource base, thereby improving system reliability as a whole." See http://www.wapa.gov/es/greennews/2002/apr8'02.htm.

[44] "An additional benefit of increased competition from renewables—and thus reduced demand for fossil fuels—could be lower prices for electricity generated from fossil fuels. Several analyses [have shown] that competition from increasing renewables could reduce natural gas prices. A comprehensive modeling project of the New England Governors' Conference found that an aggressive renewables scenario, in which renewables made up half of all new generation, would depress natural gas prices enough to lead to a slight overall reduction in regional electricity prices compared with what prices would be if new generation came primarily from fossil fuels." See Union of Concerned Scientists, *Benefits of Renewable Energy Use*, Powerful Solutions: Seven Ways to Switch America to Renewable Electricity, 1999,
http://www.ucsusa.org/clean_energy/

[45] CRS Report R40565, *Biomass Resources: The Southeastern United States and the Renewable Electricity Standard Debate*, by Richard J. Campbell.

[46] CRS Report R40631, *Water Issues of Concentrating Solar Power (CSP) Electricity in the U.S. Southwest*, by Nicole T. Carter and Richard J. Campbell.

[47] Dan Piller, *Eastern Governors Protest Midwest Wind Transmission Line*, Des Moines Register, July 30, 2010,
http://blogs.

[48] FERC recently issued its Notice of Proposed Rulemaking for "Transmission Planning and Cost Allocation by Transmission Owning and Operating Public Utilities" (FERC Docket RM10-23-000 issued June 17, 2010). FERC is seeking public comment on its intent to amend its regulations so that transmission planning would reflect "additional public policy objectives" beyond provision of reliable and cost-effective service (such as the integration of renewable energy resources).

[49] Demand-side management refers to the planning, implementation, and monitoring of utility activities designed to encourage consumers to modify patterns of electricity usage, including the timing and level of electricity demand. It refers to only energy and load-shape modifying activities that are undertaken in response to utility-administered programs. It does not refer to energy and load-shaped changes arising from the normal operation of the marketplace or from government-mandated energy-efficiency standards. Demand-side management covers the complete range of load-shape objectives, including strategic conservation and load management, as well as strategic load growth. See http://www.eia.doe.gov/glossary/index.cfm?id=D.

[50] Characteristics of a Smart Grid are described in Title XIII of the Energy Independence and Security Act of 2007. (P.L. 110-14). See http://www.ferc.gov/industries

[51] R. G. Pratt, P. J. Balducci, and C. Gerkensmeyer, et al., *The Smart Grid: An Estimation of the Energy and CO₂ Benefits*, Pacific Northwest National Laboratory, PNNL-19112, Revision 1, January 28, 2010, http://www.pnl.gov/ news/release.aspx?id=776.

[52] http://www.npr.org/templates/story/story.php?storyId=103545351.

[53] http://online.wsj.com/article/SB122722654497346099.html.

[54] CRS Report RL30153, *Critical Infrastructures: Background, Policy, and Implementation*, by John D. Moteff.

[55] Johannes Pfeifenberger, *Renewable Energy Development and Transmission Expansion – Who Benefits and Who Pays*, Brattle Group, October 12, 2010, http://www.brattle.com/_documents/UploadLibrary/Upload887.pdf.

[56] U.S. Department of Energy, *National Electric Transmission Congestion Report and Final National Corridor Designations - Frequently Asked Questions*, October 2007, http://nietc.anl.gov/documents/docs/FAQs_re_National_Corridors_10_02_07.pdf.

[57] Section 409 of ARRA (P.L. 111-5) directs DOE to analyze transmission needs and constraints related to renewable energy in the 2009 study of electric transmission congestion, and make recommendations to achieve "adequate transmission capacity."

[58] North American Electric Reliability Corporation, *Accommodating High Levels of Variable Generation*, April 2009, http://www.nerc.com/files/IVGTF_Report_041609.pdf.

[59] President Obama has declared a goal for the United States to become the world's leading exporter of renewable energy technologies, setting out policy objectives for the development of related "green jobs". *Obama Administration's Plan for Energy: An Overview*, The White House, http://www.whitehouse.gov/issues

[60] The United States is 11th among G-20 nations in clean energy investment intensity (i.e., clean energy investment as a percentage of gross domestic product). See "Who's Winning the Clean Energy Race?" at http://www.pewtrusts.org/uploadedFiles/wwwpewtrustsorg/Reports/Global_warming/G-20%20Report.pdf?n=5939.

[61] Clusters can be defined as geographic concentrations of interconnected companies, specialized suppliers, service providers, and associated institutions in a particular field that are present in a nation or region. See CRS Report R40833, *Renewable Energy—A Pathway to Green Jobs?*, by Richard J. Campbell and Linda Levine.

[62] See the report by the Breakthrough Institute and the Information Technology and Innovation Foundation "Rising Tigers, Sleeping Giant," http://thebreakthrough.org/blog/Rising_Tigers.pdf.

[63] The Advanced Projects Research Agency – Energy (ARPA-E) was established within the U.S. Department of Energy under the 2007 America COMPETES Act to promote and fund research and development of advanced energy technologies. See http://arpa-e.energy.gov/About.aspx.

[64] Center for American Progress, *Out of the Running?*, March 2010, http://www.americanprogress.org/issues pdf/out_ofjunning.pdf.

[65] David Kreutzer, Karen Campbell, and William Beach, et al., *A Renewable Electricity Standard: What It Will Really Cost Americans*, Heritage Foundation, Center for Data Analysis Report #10-03, May 5, 2010, http://www.heritage.org/ research/reports/2010/05/a-renewable-electricity-standard-what-it-will-really-cost-americans.

[66] Alternative compliance payments have been proposed as a mechanism for electric retail providers who cannot obtain sufficient renewable electricity or energy efficiency resources to meet RES compliance obligations. Portions or all of these payments are proposed to be returned to the states in an escrow or similar fund specifically to address RES issues.

[67] Between 2000 and 2009, FITs in Europe are responsible for deployment of more than 15 GW of solar PV capacity, and more than 55 GW of wind power. FITs have also led to the deployment of 75% of solar PV and 45% wind power around the world. Karlynn Cory, Toby Couture, and Claire Kreycik, *Feed-in Tariff Policy: Design, Implementation, and RPS Policy Interactions*, National Renewable Energy Laboratory, NREL/TP-6A2-45549, March 2009, http://www.nrel.gov/docs/fy09osti/45549.pdf. (NREL FIT).

[68] FITs can differentiate the tariff prices to account for different technologies, project sizes, locations, and resource intensities. FITs can also target distributed generation specifically. Toby Couture, Karlynn Cory, and Emily Williams, et al., *A Policymaker's Guide to Feed-in Tariff Policy Design*, National Renewable Energy Laboratory, NREL/TP-6A2-44849, July 2010, http://www.nrel.gov/docs/fy10osti/44849.pdf.

[69] Paul Gipe, *Washington State Introduces Feed-in Tariff*, Renewable Energy World, February 2, 2009, http://www.renewableenergyworld.com/rea/news/article/2009/02/washington-state

[70] See 132 FERC 61,047.

[71] See FERC Order 697-A,"[a]ll sales of energy or capacity made by Qualifying Facilities 20 MW or smaller are exempt from section 205 of the Federal Power Act." 18 C.F.R. 292.601.

[72] Avoided cost is the "[i]ncremental cost of alternative electric energy" in PURPA, defined "with respect to electric energy purchased from a qualifying cogenerator or qualifying small power producer, the cost to the electric utility of the electric energy which, but for the purchase from such cogenerator or small power producer, such utility would generate or purchase from another source." See 16 U.S.C. 824a-3. Cogeneration and small power production.

[73] SustainableBusiness.com News, *California Proposes Feed-in Tariff Pilot Program for Renewables*, August 27, 2010, http://www.sustainablebusiness.com/index.cfm/go/news.display/id/20942.

[74] Introduced in September 2010 by Senators Bingaman and Brownback.

[75] 16 U.S.C. §796 (17)(E).

[76] Essentially, new capacity from efficiency upgrades, new capacity, or powering of non-electric dams.

[77] For example, research is ongoing to convert carbon dioxide into methanol or other fuels. See https://share.sandia.gov/ news/resources/releases/2007/sunshine.html.

[78] National Energy Technology Laboratory, *Research Projects to Convert Captured CO_2 Emissions to Useful Products*, July 2010, http://www.netl.doe.gov/publications/press/2010/100706-Research_Projects_To_ Convert.html.

[79] See §610(a)(11).

[80] A determination of costs and benefits of the CES compared with the RES is beyond the scope of this report as it would require a number of specific assumptions to be made regarding the timing, funding, facility types and number of installations not possible from the information provided in either bill.

[81] Introduced by Senator Graham in September 2010.

[82] "Incremental Fossil Fuel Production" means the power generated at these facilities attributable to energy efficiency improvements in excess of average electric generation in the same three year period. "Incremental nuclear production" is the incremental power attributable to permanent plant energy efficiency improvements or capacity additions made on or after the date of enactment of the section. "Qualified nuclear energy" means power from a unit placed in service on or after the date of enactment of the section.

[83] P.L. 110-246.

[84] See CRS Report R40529, *Biomass: Comparison of Definitions in Legislation Through the 111th Congress*, by Kelsi Bracmort and Ross W. Gorte, for a discussion of biomass definitions.

[85] Adapted from CRS Report R40837, *Summary and Analysis of S. 1462: American Clean Energy Leadership Act of 2009, As Reported*, coordinated by Mark Holt and Gene Whitney.

In: Clean and Renewable Energy Standards   ISBN: 978-1-61324-932-1
Editors: B. J. Ruther, J. R. Moran   ©2012 Nova Science Publishers, Inc.

Chapter 3

PRESIDENT OBAMA'S PLAN TO WIN THE FUTURE BY PRODUCING MORE ELECTRICITY THROUGH CLEAN ENERGY[*]

A global race is underway to develop and manufacture clean energy technologies, and we are competing with other countries that are playing to win. America has the most dynamic economy in the world, but we can't expect to win the future by standing still. That's why, in his State of the Union address, President Obama proposed an ambitious but achievable goal of generating 80 percent of the Nation's electricity from clean energy sources by 2035. Meeting that target will position the United States as a global leader in developing and manufacturing cutting-edge clean energy technologies. It will ensure continued growth in the renewable energy sector, building on the progress made in recent years. And it will spur innovation and investment in our nation's energy infrastructure, catalyzing economic growth and creating American jobs.

---

[*] This is an edited, reformatted and augmented version of The White House publication.

- **Double the share of clean electricity in 25 years:** Currently, 40 percent of our electricity comes from clean energy sources. President Obama is calling for a national goal of doubling the share of clean energy to 80 percent by 2035.
- **Draw on a wide range of clean energy sources:** To give utilities the flexibility to generate clean energy wherever makes the most sense, all clean sources – including renewables, nuclear power, efficient natural gas, and coal with carbon capture and sequestration – would count toward the goal.
- **Deploy capital investment to sustain and create jobs:** The private sector is currently sitting on billions of dollars of capital, as investors and businesses wait to see what policies the future holds. By providing a clear signal towards a clean energy future, the President's proposal will move that capital off of the sidelines and into the economy, mobilizing tens of billions of dollars each year in new investment and creating jobs across the country.
- **Drive innovation in clean energy technologies:** The engine of economic strength is technological innovation. By providing American businesses a market here at home for innovative clean energy technologies, we will unleash the creative power of American entrepreneurs – and ensure that our nation leads the world in clean energy.
- **Complement the clean energy research and development agenda:** The President's Budget proposes to increase overall investment in clean energy technologies by about one-third compared to 2010, including doubling energy efficiency investment at the Department of Energy and increasing investment in the Advanced Research Projects Agency-Energy (ARPA-E) program to push bold new ideas through to commercialization.

THE PRESIDENT'S PROPOSAL FOR A CLEAN ENERGY STANDARD

The President is proposing that a new Clean Energy Standard (CES) be founded on five core principles:

- Doubling the share of clean electricity over the next 25 years. To mobilize capital and provide a strong signal for innovation in the energy sector, a CES should be established that steadily increases the share of delivered electricity generated from clean energy sources, rising from 40 percent today to 80 percent by 2035.

- Credit a broad range of clean energy sources. To ensure broad deployment and provide maximum flexibility in meeting the target, clean energy credits should be issued for electricity generated from renewable and nuclear power; with partial credits given for clean coal and efficient natural gas.
- Protecting consumers against rising energy bills. The CES should be tailored to protect consumers, and coupled with smart policies that will help American families and businesses save money by saving energy.
  - o The CES should be paired with energy efficiency programs that will lower consumers' energy bills, such as stronger appliance efficiency standards, tax credits for energy efficiency upgrades, and the proposed Home Star program.
  - o The CES should also include provisions to help manufacturers invest in technologies to improve efficiency and reduce energy costs.
- **Ensuring fairness among regions.** Different regions of the country rely on diverse energy sources today, and have varying clean energy resources for the future. The CES must ensure that these differences are taken into account – both among regions and between rural and urban areas.
- **Promoting new technologies such as clean coal.** The CES should include provisions to encourage deployment of new and emerging clean energy technologies, such as coal with carbon capture and sequestration.

BUILDING ON PROGRESS

An ambitious Clean Energy Standard will build on the enormous recent progress in renewable energy, with 16,000 megawatts of new electric generating capacity from wind, solar, and geothermal energy that has come online since 2008 – an increase of nearly 60 percent in just two years.

- **Creating jobs and clean energy through the Recovery Act:** ARRA made an historic investment in clean energy of over $90 billion, which has already created or saved 224,500 American jobs and tens of thousands of domestic renewable energy projects – including some of the largest in the world – putting the country on target to double renewable energy generation by 2012.

- **Expanding production through the successful "1603" grant program:** The renewable energy grant program under the Recovery Act has been an essential tool in deploying renewable energy resources in the U.S. over the past two years, successfully increasing U.S. manufacturing and creating tens of thousands of new American jobs. The Recovery Act converted these pre-existing tax credits into grant payments, making it easier for recipients to quickly expand clean energy generation and hiring. To date, the 1603 program has helped encourage more than 4,000 clean energy projects. The Tax Relief, Unemployment Insurance Reauthorization, and Job Creation Act of 2010 extended the 1603 program for one year.

- **Staying on the cutting edge through Clean Energy R&D:** Through the Recovery Act, we have invested in 120 cutting edge research projects through Advanced Research Projects Agency – Energy (ARPA-E) program across areas ranging from grid technology and power electrics to nuclear technology and batteries and energy storage. Past Budgets funded three "Energy Innovation Hubs" that explore building efficiency, fuel from sunlight, and nuclear reactor modeling and simulation. This year's Budget will more than double funding for ARPA-E and will double the number of Hubs.

- **Siting a record number of renewable projects on public lands:** In the last year alone, the Department of Interior green-lighted the first nine commercial-scale solar energy projects for construction on public lands, including the largest solar power plants in the world. When built, these projects will supply nearly 3,700 MW of power—enough to power more than 1,100,000 homes—and are expected to create about 7,300 new jobs. Interior also approved the first offshore wind farm in the U.S., and has launched an initiative to accelerate the rapid and responsible development of America's vast offshore wind resources.

In: Clean and Renewable Energy Standards ISBN: 978-1-61324-932-1
Editors: B. J. Ruther, J. R. Moran ©2012 Nova Science Publishers, Inc.

Chapter 4

WHITE PAPER ON A CLEAN ENERGY STANDARD[*]

Sens. Jeff Bingaman and Lisa Murkowski

PURPOSE

In his recent State of the Union address, President Obama proposed a Clean Energy Standard (CES) to require that 80 percent of the nation's electricity come from clean energy technologies by 2035. The Senate Energy and Natural Resources (ENR) Committee now faces a threshold question of what the general policy goals for the electric sector are and whether a CES would most effectively achieve them. Is the goal to reduce greenhouse gas emissions, lower electricity costs, spur utilization of particular assets, diversify supply, or some combination thereof? Depending on the goals, is a CES the right policy for the nation at this time? If so, is 80 percent by 2035 the right target? If not, should alternatives to reach similar goals be considered?

The purpose of this document is to lay out some of the key questions and potential design elements of a CES, in order to solicit input from a broad range of interested parties, to facilitate discussion, and to ascertain whether or not consensus can be achieved.

---

[*] This is an edited, reformatted and augmented version of a Committee on Energy and Natural Resources, United States Senate publication, dated March 21, 2011.

INTRODUCTION

Advocates of a CES maintain that requiring the deployment of increasing amounts of clean electricity can lead to a variety of benefits, such as the reduction of greenhouse gases and other emissions, as well as an increase in domestic manufacturing of associated technologies. In contrast, opponents have claimed that a federal electricity mandate, depending on its design, could pick winners and losers among competing technologies and serve as a tax that may cause a wealth transfer from those regions of the country lacking compliant resources.

Congress has debated Renewable Portfolio and Renewable Electricity Standards (RES) for the past decade. During the 111th Congress, the ENR Committee included a 15 percent by 2021 RES in S. 1462, the American Clean Energy Leadership Act of 2009. While a number of CES proposals have been introduced or discussed in past Congresses, the concept has not yet been seriously considered or analyzed.

Over time, there have been a variety of goals advanced for deployment of clean or renewable energy. For some, the primary focus of an RES has been to enhance the competitiveness of renewable technologies in the short term, in order to allow them to become economically competitive with fossil technologies. Other proposals have focused on diversifying electric generation in order to guard against possible resource constraints. Still others highlight the emissions reduction potential of these technologies.

If the ENR Committee elects to develop a CES, there are a number of design questions that require careful consideration. The decisions made in the design of such a standard will necessarily favor certain priorities over others.

CURRENT STATE OF CLEAN ENERGY DEPLOYMENT

Data from the Energy Information Administration (EIA) indicates that in 2010, domestic electricity generation was comprised of about 20 percent from nuclear power plants, 10 percent from renewable energy power plants (hydropower, wind, solar, geothermal and biomass), 25 percent from natural gas power plants, and 45 percent from coal power plants. If clean energy were defined as renewable and nuclear energy only, then the United States would currently be obtaining 30 percent of its electricity from clean sources. If efficient natural gas (i.e. combined cycle) were included as well – and awarded "half credits" in accordance with the President's CES proposal – the United States would currently be obtaining 40 percent of its electricity from clean sources.

The EIA reference scenario, in its 2011 Annual Energy Outlook, projects that overall electric generation will increase by about 20 percent between 2010 and 2035. The majority of new capacity is expected to come from natural gas power plants. Natural gas is expected to maintain its 25 percent share of overall electricity generation throughout this period. Renewable power is expected to grow to a 14 percent share of the generation mix. Nuclear is expected to add capacity but decrease slightly in its overall share of the generation mix to 17 percent in 2035. Events in Japan may affect that potential growth in capacity. Generation from coal is expected to increase overall but decrease to a 42 percent share of the generation mix.

KEY ELEMENTS FOR CLEAN ENERGY STANDARD PROPOSALS

1. What Should Be the Threshold for Inclusion in the New Program?

In the RES contained in S. 1462 last Congress, utilities selling four million megawatt hours or more of retail electric power in a calendar year would have been subject to the mandate. Additionally, the State of Hawaii was specifically excluded from the program's requirements. The President's CES proposal does not appear to contain a threshold for inclusion, which means that all electric utilities, regardless of size, would be responsible for meeting any new requirements imposed by a CES.

Key Questions
- Should there be a threshold for inclusion or should all electric utilities be subject to the standards set by a CES?
- Should any states or portions of states be specifically excluded from the new program's requirements?
- How should a federal mandate interact with the 30 existing state electricity standards?

2. What Resources Should Qualify as "Clean Energy"?

The definition of what qualifies as "clean energy" will be crucial in determining the overall mix of technologies deployed to comply with a CES. While previous CES proposals have gone beyond the narrow set of renewable technologies allowed under a RES, by including nuclear plants and coal plants with carbon capture and storage (CCS), the President's proposal also seeks to allow efficient natural gas without CCS to count towards compliance. While past proposals have credited energy efficiency measures to varying degrees, the President's CES proposal does not give clean energy credits for energy efficiency measures.

Key Questions
- On what basis should qualifying "clean energy" resources be defined? Should the definition of "clean energy" account only for the greenhouse gas emissions of electric generation, or should other environmental issues be accounted for (e.g. particulate matter from biomass combustion, spent fuel from nuclear power, or land use changes for solar panels or wind, etc.).
- Should qualifying clean energy resources be expressly listed or based on a general emissions threshold? If it is determined that a list of clean energy resources is preferable, what is the optimal definition for "clean energy" that will deploy a diverse set of clean generation technologies at least cost? Should there be an avenue to qualify additional clean energy resources in the future, based on technological advancements?
- What is the role for energy efficiency in the standard? If energy efficiency qualifies, should it be limited to the supply side, the demand side, or both? How should measurement and verification issues be handled?
- Should retrofits or retirements of traditional fossil-fuel plants be included in the standard?
- Should the standard be focused solely on electricity generation, or is there a role for other clean energy technologies that could displace electricity, such as biomass-to thermal energy?

3. How Should the Crediting System and Timetables Be Designed?

The design of the crediting system and the timing and stringency of the targets will necessarily impact the mix of technologies deployed as well as the ultimate costs imposed on end-use customers. For example, previous RES and CES proposals have called for taking certain existing technologies out of the baseline for purposes of calculating the mandate (e.g. conventional hydropower), while providing full credits to new resources.

Key Questions
- Should the standard's requirements be keyed to the year 2035 or some other timeframe?
- What interim targets and timetables should be established to meet the standard's requirements?
- What are the tradeoffs between crediting all existing clean technologies versus only allowing new and incremental upgrades to qualify for credits? Is one methodology preferable to the other?
- Should partial credits be given for certain technologies, like efficient natural gas and clean coal, as the President has proposed? If partial credits are used, on what basis should the percentage of credit be awarded? Should this be made modifiable over the life of the program?
- Is there a deployment path that will optimize the trade-off between the overall cost of the program and the overall amount of clean energy deployed?
- What would be the effect of including tiers for particular classes of technology, or for technologies with different levels of economic risk, and what would be a viable way of including such tiers?
- Should the same credit be available to meet both the federal mandate and an existing state standard or should a credit only be utilized once?
- Should there be a banking and/or borrowing system available for credits and, if so, for how long?

4. How Will a CES Affect the Deployment of Specific Technologies?

The value and expected future value of clean energy credits created by a CES will be the primary driver of clean energy deployment. Each technology faces different economic and financing issues. Some, such as nuclear energy, face significant upfront capital costs but low ongoing fuel costs. Others, such as natural gas power plants, may be deployed relatively inexpensively but with a higher percentage of ongoing costs coming from fuel. How credit value changes the economics of each individual technology will determine which technologies get deployed.

Key Questions
- How valuable would clean energy credits have to be in order to facilitate the deployment of individual qualified technologies?
- How might a CES alter the current dispatch order of existing generation (such as natural gas-fired power plants), which has been driven by minimization of consumer costs, historically?
- What is the expected electricity generation mix for a target of 80 percent clean energy by 2035, under the President's proposal or an alternative construct?
- Could different crediting and requirements than those proposed by the President be more effective in deploying clean technologies?

5. How Should Alternative Compliance Payments, Regional Costs, and Consumer Protections Be Addressed?

In considering a CES, it is important to consider the additional costs that may accompany such a policy and how those costs may vary by region. Some regions of the country contain more abundant energy resources than others, and utilities within those regions may be utilizing vastly different fuel mixes. Important design goals for a CES are to ensure price certainty for consumers and industry, minimize regional disparities in the cost of such a policy, and contain costs overall. The RES contained in S. 1462 last Congress included cost containment mechanisms such as limiting the electric rate impact of a utility's incremental compliance costs to not more than four percent per retail customer annually; an Alternative Compliance Payment (ACP) that was available for utilities that determined they could not meet the program requirements; providing a potential variance if transmission constraints prevent service delivery; and potentially allowing waivers for reasons of Force Majeure.

Key Questions
- What are the anticipated effects on state and regional electricity prices of a CES structured according to the President's proposal? What are the anticipated net economic effects by region?
- Would other CES formulations or alternative policy proposals to meet a comparable level of clean energy deployment have better regional or net economic outcomes?
- How might various price levels for the ACP affect the deployment of clean energy technologies?
- What options are available to mitigate regional disparities and contain costs of the policy?
- What are the possible uses for potential ACP revenues? Should such revenues be used to support compliance with the standard's requirements? Should all or a portion of the collected ACP revenues go back to the state from which they were collected? Should ACP revenues be used to mitigate any increased electricity costs to the consumer that may be associated with the CES?
- Should cost containment measures and other consumer price protections be included in a CES?
- How much new transmission will be needed to meet a CES along the lines of the President's proposal and how should those transmission costs be allocated?
- Are there any technological impediments to the addition of significantly increased renewable electricity generation into the electrical grid?
- What are the costs associated with replacing or retrofitting certain assets within the existing generation fleet in order to meet a CES?
- What level of asset retirements from within the existing generation fleet are anticipated as a result of a CES?

6. How Would the CES Interact with Other Policies?

The credit value generated by imposition of a CES may not, by itself, be enough to address obstacles faced by particular clean energy technologies. For example, the deployment of solar panels has raised concerns about land use changes in certain desert areas. Coal with CCS confronts post-closure liability issues and the extraction of the feedstock itself has become subject to increasingly stringent regulatory treatment. For nuclear power, financing new projects has been difficult due to significant, up-front capital costs. All domestic energy development projects face substantial permitting hurdles. Reaching the President's CES target of 80 percent by 2035 will require a diverse set of resources, so technology-specific supporting policies may be necessary.

Key Questions

- To what extent does a CES contribute to the overall climate change policy of the United States, and would enactment of a CES warrant changes to other, relevant statutes?
- What are the specific challenges facing individual technologies such as nuclear, natural gas, CCS, on- and offshore wind, solar, efficiency, biomass, and others?
- Will the enactment of a CES be sufficient for each technology to overcome its individual challenges?
- Should there be an examination of energy connected permitting?
- Are there specific supporting policy options that should be considered for coal, nuclear, natural gas, renewable energy, and efficiency?
- What is the current status of clean energy technology manufacturing, and is it reasonable to expect domestic economic growth in that sector as a result of a CES?

In: Clean and Renewable Energy Standards   ISBN: 978-1-61324-932-1
Editors: B. J. Ruther, J. R. Moran   ©2012 Nova Science Publishers, Inc.

Chapter 5

RENEWABLE PORTFOLIO STANDARDS (RPS) AND GOALS[*]

A Renewable Portfolio Standard (RPS) or Energy Standard (RES) requires a percent of energy sales (MWh) or installed capacity (MW) to come from renewable resources. Percents usually increase incrementally from a base year to a later target. The map on the front shows ultimate targets.

- Renewable energy certificates or credits (RECs) allow states to verify RPS compliance. Utilities may comply either by generating or purchasing renewables under contract, or by buying RECs. Most states set an alternative compliance payment (ACP) for shortfalls. Each state's ACP is different.
- 19 states plus D.C. have renewable mandates. Seven have renewable goals without financial penalties. Nebraska's two largest public power districts have renewable goals. The Tennessee Valley Authority has a goal for its 7-state region.
- In 2010, one state enacted a renewable goal. Two states introduced RPS legislation that did not pass; 2 others saw bills that proposed increased targets fail. 3 states increased or extended existing renewable targets; 2 expanded the definition of eligible resources. One state passed major legislation that will allow the growth of renewable capacity for export, and another began a renewable pilot.
- Sixteen states plus D.C. have final targets of 20% or higher. Three of these set lower targets for cooperatives (co-ops) or municipal electric utilities (munis).

---

[*] This is an edited, reformatted and augmented version of a Renewable Power & Energy Efficiency Market: Renewable Portfolio Standards publication, from www.ferc.gov/oversight, dated August 11, 2010.

Final Target	Number	States with Renewable Mandates (RPS)
10% - 14%	6	Iowa, Mich., N.C., Ohio, Texas, Wis.
15% - 19%	7	Ariz., Mass., Mo., Mont., Pa., R.I., Wash.
20%	4	D.C., Kansas, Md., N.M.
21% - 24%	2	N.H., N.J.
25% - 29%	6	Conn., Del., Ill., Minn., Nev., Ore. Calif., Colo.,
30% - 39%	3	N.Y.
40%	2	Hawaii, Maine

State Renewable Actions:

- Delaware extended and increased its RPS to 25% by 2025, from 20% by 2019. It also increased the solar photovoltaic (PV) carve-out to 3.5% by 2025 from 2% by 2019. (July 28)
- Massachusetts revised its definition of eligible biomass to assure that its RPS comports with its greenhouse gas reduction targets. It will allow sustainably harvested biomass used in highly efficient combined heat and power (CHP) facilities. It set 60% as the minimum efficiency, with a goal of 80% efficiency by 2020. (July 7)
- Louisiana's PSC voted unanimously to implement a pilot renewable program to assess costs and benefits of a potential RPS. The pilot has two parts: a small-scale technology research program and a 350 MW utility capacity target. Utilities must issue requests for renewable resources in proportion to their percent of retail sales. Capacity must be able to be online by 2012-2013. (June 23)
- Texas opened a rulemaking on an RPS modification that would create a mandate for renewable technologies, including solar, now covered by a 500 MW non-wind goal. The PUC held hearings and public workshops to examine costs and benefits of proposed changes and to consider other states' experiences with solar programs. The PUC expects to reach a decision in the latter half of 2010.
- 16 states have solar or distributed generation (DG) set-asides in their RPS to encourage development of these higher-cost technologies so they can move closer to cost-parity with other resources. Three states award renewable credit multipliers for solar or DG; 5 states use set-asides and multipliers. 13 states adopted or increased solar targets in an RPS in the last two years; 5 of those were in 2010. Details available at: http://www.ferc.gov/marketoversight/othr-mkts/renew/othr-rnw-rps-solar-DG.pdf

29 states and D.C. have an RPS; 7 States and 3 Power Authorities have Goals

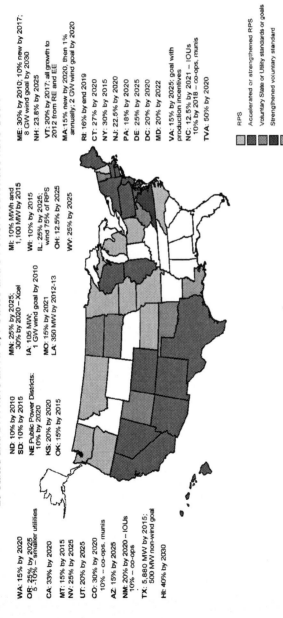

WA: 15% by 2020

OR: 25% by 2025;
5–10% – smaller utilities

CA: 33% by 2020

MT: 15% by 2015
NV: 25% by 2025

UT: 20% by 2025

CO: 30% by 2020
10% – co-ops, munis

AZ: 15% by 2025

NM: 20% by 2020 – IOUs
10% – co-ops

TX: 5,880 MW by 2015;
500 MW non-wind goal

HI: 40% by 2030

ND: 10% by 2010
SD: 10% by 2015

NE Public Power Districts:
10% by 2020

KS: 20% by 2020

OK: 15% by 2015

MN: 25% by 2025;
30% by 2020 – Xcel

IA: 105 MW;
1 GW wind goal by 2010

MO: 15% by 2021

LA: 350 MW by 2012-13

MI: 10% MWh and
1,100 MW by 2015

WI: 10% by 2015

IL: 25% by 2025;
wind 75% of RPS

OH: 12.5% by 2025

WV: 25% by 2025

ME: 30% by 2010; 10% new by 2017;
8 GW wind goal by 2030

NH: 23.8% by 2025

VT: 20% by 2017; all growth to
2012 from RE and EE

MA: 15% new by 2020, then 1%
annually; 2 GW wind goal by 2020

RI: 16% by end 2019

CT: 27% by 2020

NY: 30% by 2015

NJ: 22.5% by 2020

PA: 18% by 2020

DE: 25% by 2025

DC: 20% by 2020

MD: 20% by 2022

VA: 15% by 2025; goal with
production incentives

NC: 12.5% by 2021 – IOUs
10% by 2018 – co-ops, munis

TVA: 50% by 2020

Legend:
- RPS
- Accelerated or strengthened RPS
- Voluntary State or Utility standards or goals
- Strengthened voluntary standard
- Pilot or study

Updates at: http://www.ferc.gov/market-oversight

Notes: A RPS requires a percent of an electric provider's energy sales (MWh) or installed capacity (MW) to come from renewable resources. Most specify sales (MWh). Map percents are final years' targets.
Nebraska's two largest public power districts, which serve close to two-thirds of Nebraska load, have renewable goals. The Tennessee Valley Authority's (TVA) goal across its 7-state territory is 50% zero- or low-carbon generation by 2020.

Sources: derived from data in: Lawrence Berkeley Labs, State Public Utility Commission (PUC) and legislative tracking services, Pew Center. Details, including timelines, are in the Database of State Incentives for Renewables and Energy Efficiency: http://www.dsireusa.org

In: Clean and Renewable Energy Standards   ISBN: 978-1-61324-932-1
Editors: B. J. Ruther, J. R. Moran   ©2012 Nova Science Publishers, Inc.

Chapter 6

RENEWABLE PORTFOLIO PROVISIONS FOR SOLAR AND DISTRIBUTED GENERATION[*]

Policies incent solar and DG development:

- 16 states and D.C. have created solar or distributed generation (DG) set-asides in their Renewable Portfolio Standards (RPS), to encourage development of higher-cost technologies so they can move closer to cost parity with other renewable resources.
- Three states use solar credit multipliers instead of set-asides. Five use both set-asides and multipliers. Two states with renewable energy goals provide solar incentives.
- Set-asides specify what portion of an RPS should come from a specific technology. Multipliers increase the value of renewable energy certificates (RECs) awarded for each MWh produced by eligible technologies. Some states have separate solar RECs (SRECs) that can be traded.
- 12 states and D.C. added or increased solar policies in an RPS in the last two years.
- Two states have solar targets or programs outside an RPS or renewable goal: California and Rhode Island.

---

[*] This is an edited, reformatted and augmented version of a Renewable Power & Energy Efficiency Market: Solar and Distributed Generation RPS provisions publication, from www.ferc.gov/oversight., dated August 6, 2010.

- Lawrence Berkeley (LNBL) projected that existing solar carve-outs require 560 MW of solar through 2010 and 8,447 MW by 2025. That development is exclusive of non-RPS goals, such as California's "million roofs" program. LBNL found that multipliers have been less effective in stimulating solar development than set-asides.
- Feed-in Tariffs (FIT), or advanced renewable incentives, require that utilities buy qualified renewable generation at a fixed rate, higher than that provided to other generators, under multi-year contracts. FITs were created or enhanced by five states and two munis in 2009: California, Hawaii, Oregon, Vermont, Washington, Gainesville (FL) and Sacramento (SMUD), CA.

Recent State Activities:

- Texas has an open rulemaking on whether to modify its RPS to create a mandate for renewable technologies, including solar, now covered by a 500 MW non-wind goal. The PUC held a hearing and public workshop to look at costs and benefits of proposed changes and to consider the experience of other states with solar programs. (June 11)
- Massachusetts issued final regulations for its Solar Carve-Out program, applicable to all retail suppliers. Eligible customer-sited projects up to 2 MW must be in-state, except projects under contract prior to January 2011. MA will hold a clearinghouse auction for surplus SRECs until its 400 MW solar target is met. (June 8)
- Missouri's PSC adopted regulations for the RPS passed by ballot in Nov 2008. IOUs can comply with the 15% by 2021 overall target either by generating electricity from renewables or buying RECs. The 2% solar carve-out, however, requires that utilities use standard-offer contracts for solar RECs, and prohibits SREC purchases by IOUs from affiliates. (June 3)
- California has a 3,000 MW solar goal beyond its RPS. The California Solar Initiative (CSI) is the largest of its four solar programs. The CSI is a performance-based incentive program with a goal of 1,940 MW new PV by 2016 in three IOU territories. 366 MW were installed under CSI through July.

16 States and D.C. use Set-asides, 3 use Multipliers to Encourage these Technologies

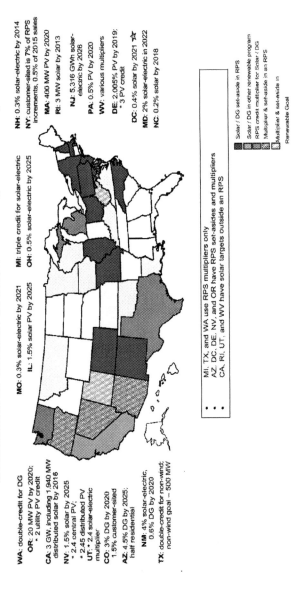

WA: double-credit for DG

OR: 20 MW PV by 2020;
* 2 utility PV credit

CA: 3 GW, including 1,940 MW
distributed solar by 2016

NV: 1.5% solar by 2025
* 2.4 central PV;
* 2.45 distributed PV
UT: * 2.4 solar-electric
multiplier

CO: 3% DG by 2020
1.5% customer-sited

AZ: 4.5% DG by 2025;
half residential

NM: 4% solar-electric,
0.6% DG by 2020

TX: double-credit for non-wind;
non-wind goal – 500 MW

MO: 0.3% solar-electric by 2021

IL: 1.5% solar PV by 2025

MI: triple credit for solar-electric

OH: 0.5% solar-electric by 2025

NH: 0.3% solar-electric by 2014

NY: customer-sited is 7% of RPS
increments, 0.5% of 2015 sales

MA: 400 MW PV by 2020

RI: 3 MW solar by 2013

NJ: 5,316 GWh solar-
electric by 2026

PA: 0.5% PV by 2020

WV: various multipliers

DE: 2.005% PV by 2019;
* 3 PV credit

DC: 0.4% solar by 2021 ☆

MD: 2% solar-electric in 2022

NC: 0.2% solar by 2018

• • •
MI, TX, and WA use RPS multipliers only
AZ, DC, DE, NV, and OR have RPS set-asides and multipliers
CA, RI, UT, and WV have solar targets outside an RPS

Solar / DG set-aside in RPS

Solar / DG in other renewable program

RPS credit multiplier for Solar / DG

Multiplier & set-aside in an RPS

Renewable Goal

Updates at: http://www.ferc.gov/market-oversight

Notes: (*) Multipliers receive extra credit towards RPS compliance. Set-asides are specific technology targets in an RPS, specified by percent, MW, or MWh. An RPS requires a percent of an electric provider's energy sales (MWh) or installed capacity (MW) to come from renewable resources.

Abbreviations: DG – distributed generation; PV – solar photo-voltaic; RPS – Renewable Portfolio Standard

Sources: Derived from data in: LBNL, State Legislative and Public Utility web sites, California Solar Initiative, and the Database of State

Incentives for Renewables and Energy Efficiency: http://www.dsireusa.org.

INDEX